Predictive Microbiology:
Theory and Application

INNOVATION IN MICROBIOLOGY SERIES

Series Editor: **Dr A. N. Sharpe,**
Health and Welfare, Ottawa, Canada

Titles available:

2. Utilization of Microorganisms in Meat Processing: A handbook for meat plant operators
   Jim Bacus

4. Membrane Filter Food Microbiology
   Anthony N. Sharpe *and* **Pearl I. Peterkin**

5. Predictive Microbiology: Theory and Application
   T. A. McMeekin, J. N. Olley, T. Ross *and* **D. A. Ratkowsky**

Predictive Microbiology: Theory and Application

T. A. McMeekin
University of Tasmania, Australia

J. N. Olley
University of Tasmania, Australia

T. Ross
University of Tasmania, Australia

and

D. A. Ratkowsky
Department of Primary Industries and Fisheries, Tasmania, Australia

RESEARCH STUDIES PRESS LTD.
Taunton, Somerset, England

JOHN WILEY & SONS INC.
New York · Chichester · Toronto · Brisbane · Singapore

RESEARCH STUDIES PRESS LTD.
24 Belvedere Road, Taunton, Somerset, England TA1 1HD

Marketing and Distribution:

Australia and New Zealand:
JACARANDA WILEY LTD.
GPO Box 859, Brisbane, Queensland 4001, Australia

Canada:
JOHN WILEY & SONS CANADA LIMITED
22 Worcester Road, Rexdale, Ontario, Canada

Europe, Africa, Middle East and Japan:
JOHN WILEY & SONS LIMITED
Baffins Lane, Chichester, West Sussex, England

North and South America:
JOHN WILEY & SONS INC.
605 Third Avenue, New York, NY 10158, USA

South East Asia:
JOHN WILEY & SONS (SEA) PTE LTD
37 Jalan Pemimpin #05-04
Block B Union Industrial Building, Singapore 2057

Library of Congress Cataloging in Publication Data

Available.

British Library Cataloguing in Publication Data

A catalogue record for this book is available from the British Library.

ISBN 0 86380 132 3 (Research Studies Press Ltd.)
ISBN 0 471 93545 X (John Wiley & Sons Inc.)

Printed in Great Britain by SRP Ltd., Exeter

ACKNOWLEDGEMENTS

This book might be subtitled *"A Tale From Tasmania"* as it represents the collective efforts of four authors from very different backgrounds who coincidentally became resident in Tasmania.

June Olley arrived from England via Aberdeen; David Ratkowsky was originally from New York; my intention was to gain a few years experience abroad before returning to Belfast, and Tom Ross was actually born in Tasmania. Nevertheless, he too has immediate old world connections as his parents emigrated from Germany.

Given this potentially explosive ethnic mix, it is not surprising that there have been some memorable moments during the evolution of the text. It would be inappropriate to recount them here, but they do make excellent introductions to conference presentations and after dinner conversation.

When this diversity was coupled with a similar breadth in training and experience, the combination resulted in a considerable degree of synergy which has led to the completion of this text.

Readers should be aware that there is no "senior" author - the order was determined alphabetically and by institution. However, the older authors wish to make special mention of Tom Ross, a PhD student, whose practical and intellectual contributions provided the impetus to continue and complete the task. In addition to almost sole responsibility for Chapters 2 and 9, he made significant contributions to Chapter 10 and several other chapters. In agreeing to prepare the camera ready copy he also became the final arbiter for inclusion or deletion of material, often returning contributions to the other authors for clarification. This painstaking and difficult role was handled with skill and, on occasions, diplomacy.

The authors are, of course, indebted to many people and organisations for their direct and indirect contributions. Funds for our research in predictive microbiology have been obtained from the Australian Research Council; University of Tasmania; CSIRO; Department of Primary Industry and Fisheries, Tasmania; Department of Industry, Trade and Commerce; Chicken Meat Research and Development Corporation; Fishing Industry Research and Development Corporation; Dairy Research and Development Corporation; Pig Research and Development Corporation; Meat Research Corporation, and the Rural

Industry Research and Development Corporation. The latter group also provided a postgraduate scholarship for Tom Ross.

We have also gained considerable benefit from collaboration and correspondence with the rapidly expanding worldwide predictive microbiology group, including Terry Roberts, Bob Buchanan, Martin Cole, Marcel Zwietering and their colleagues. Information and advice on specific topics was provided by Dr. P. Doe, University of Tasmania, Hobart; Dr. C. Jason, Aberdeen, and Professor R. Jaenicke, Regensburg.

We sincerely thank the series editor, Dr. A.N. Sharpe, for the opportunity to contribute to the IMRS series and for his considerable patience and quiet encouragement during several "editions" of the monograph.

A series of Honours and postgraduate students at the University of Tasmania have contributed to our knowledge and data base in predictive microbiology and we thank Hashim Bin Daud, Andrew Ball, Rob Chandler, Morag Anderson, Laura Hayward, Karina Neumeyer and Laura Kamperman. Innumerable early drafts of the book were typed by Sally Jones, who also proof read the final draft. Equipment repairs and laboratory requisites were attended to by Bill Peterson.

Finally, a sincere acknowledgement to the support and encouragement provided by our respective spouses - Jenny McMeekin, Frank Cumbrae-Stewart, Wendy Ross and Ann Ratkowsky. Each is secretly hoping for an end to late night and early morning phone calls describing progress on the meaning of life, the universe and everything, which is likely to be described as the *square root* of Douglas Adams' 42.

T. A. McMeekin
Hobart
November, 1992

CONTENTS

3. *MODELLING TEMPERATURE EFFECTS*

4. *COMPARING BĚLEHRÁDEK-TYPE AND ARRHENIUS-TYPE MODELS USING REAL DATA*

5. MODELLING THE COMBINED EFFECT OF TEMPERATURE, WATER ACTIVITY, AND OTHER FACTORS ON MICROBIAL GROWTH RATE

6. PREDICTING THE SHELF LIFE OF CHILL STORED FOODS BY TEMPERATURE FUNCTION INTEGRATION

7. EVALUATION OF HYGIENE AND MICROBIOLOGICAL SAFETY OF PRODUCTS AND PROCESSES

8. TIME-TEMPERATURE INDICATORS

9. ELECTRONIC TEMPERATURE FUNCTION INTEGRATION

10. A THERMODYNAMIC APPROACH TO BACTERIAL GROWTH

Professor MUDr. Jan Bělehrádek, in rectorial robes, whilst at Charles University, Prague.

PREFACE

A TRIBUTE TO PROFESSOR MUDR. JAN BĚLEHRÁDEK

There have been many attempts to predict the rate of reactions and their response to temperature over the last one hundred years. The two chief proponents have been Svante Arrhenius and Jan Bělehrádek. Arrhenius is the better known of the two men, studying at the University of Uppsala, Sweden. He is best known for his work on the ionic theory of solution of salts for which he won the Nobel Prize in 1903. Bělehrádek is less well known, his major work "Temperature and Living Matter" having been published in Berlin in 1935 in the turbulent years before the Second World War. His biographical details and photograph have been supplied kindly to the authors from the Charles University Archive Office by the Czechoslovak Academy of Sciences and are shown in Table P.1. It is tempting to think that, had he survived another two years, he might have been heartened to find his equations relating biological rates to temperature rediscovered by an entirely different route.

Both Arrhenius and Bělehrádek thought widely, Arrhenius being an early writer on the greenhouse effect. Bělehrádek's wider interests are shown in Table P.2.

His Czech biography ends in 1948. In 1954, we find his paper on "Temperature and rate of enzyme action" published in *Nature* from a private address in Paris. In 1956, he completed his review of the literature on the general physiological aspects of the action of temperature which had appeared since the publication of his book in 1935. This review, "Physiological aspects of heat and cold", was invited by the *Annual Reviews of Physiology* and published in 1959. The address given is the Middlesex Hospital Medical School, London.

Why, we must ask, have microbiologists so instinctively turned to Arrhenius kinetics and ignored the thoughts of Bělehrádek?

The Arrhenius equation (Arrhenius, 1889), as it came to be known, was derived to describe the effects of temperature on the reaction velocity of gases at infinite dilutions and was itself derived from the

Table P. 1 **Biographical Details of Bělehrádek's Early Life**

Professor MUDr. Jan Bělehrádek

Born: 18 December 1896 (Praha, Czechoslovakia)

Died: 8 May 1980 (London, United Kingdom)

Education and positions

1903-1907	Elementary school (five classes)
1908-1916	Real gymnasium (secondary modern school), eight classes (Praha)
1916-1922	Charles University, Praha (Faculty of Medicine and Faculty of Sciences)
1922	Graduated as MUDr. (Medicinae Universae Doctor - MD)
1922-1923	Post-doctoral Fellowship at the University of London
1923-1925	Assistant at the Institute of Biology, Faculty of Medicine, Masaryk's University, Brno
1925 (18 May)	Docent habilitation for the branch of general biology at the Faculty of Medicine, Masaryk's University, Brno
1925-1927	Docent (Assistant Professor) of General Biology, Faculty of Medicine, Masaryk's University, Brno
1927(20 Dec.)	Inaugurated as Extraordinary Professor of General Biology at the same University and Faculty, Brno
1931(30 June)	Inaugurated as Ordinary Professor of General Biology at the same University and Faculty, Brno
1927-1934	Head of the Institute of Biology at the same University and Faculty, Brno
1934(22 Nov.)	Nominated as Ordinary Professor of General Biology and Head of Institute of Biology, Faculty of Medicine, Charles University, Praha
1934-1939	Worked in the above mentioned position at the Charles University, Praha
1937-1938	Appointed as Decanus of the Faculty of Medicine, Charles University, Praha
1945-1948	Ordinary Professor of General Biology and Head of the Institute of Biology, Faculty of Medicine, Charles University, Praha
1945-1946	Rector of the Charles University, Praha
1946	Nominated as Doctor Honoris Causa of the Sorbonne University, France
1948	Moved abroad, worked at UNESCO in Paris and later in England (London, Plymouth, etc.)

Table P. 2 **Bělehrádek's Intellectual Interests**

Scientific Work

- Studies in comparative physiology of the pericardium
- Described the regulatory influence of temperature upon physiological processes (monograph "Temperature and Living Matter", 1935)
- Author and co-author of several text books on general biology
- Discovered the inducing effect of catabolites on anabolical processes
- Studies upon the life-role of water, the significance of various life-rhythms, the evolutionary aspects of endocrine secretion, the sexuological, genetical and eugenical topics
- Supported the application of scientific knowledge in public life and the introduction of the holistic philosophy into the sphere of science and culture (several monographs in Czech: "On Holism", 1941; "The View-Point of Biology", 1942; "The Man in Numerical Data",1945; "The New Breath", 1946; "Science and Society", 1946)

Other Activities

- Editor or co-editor of various Czech journals ("Czech Cosmos", 1922-24; "Science and Life", etc.) and encyclopaedias
- Significant activities in Czechoslovak politics and culture (till 1948), after 1945 elected as delegate to the Czechoslovak National Assembly. Chevalier de la Légion de l'Honneur (France, 1937)

equation of van't Hoff (1884) (cf. Glasstone, 1960). Later, it was used with success to describe simple chemical reactions. Arrhenius (1915) saw no reason to presuppose any difference in the laws governing 'vital' processes and those found in general chemistry. Where anomalies occurred, they could often be related to changes in pH in poorly buffered systems. In general, he regarded many life processes as first order reactions with distinctive 'Temperature Characteristics' (apparent activation energies) obtained by plotting the logarithm of the rate against the reciprocal of the absolute temperature.

Bělehrádek (1935) quoted references dating back to 1911 to show that the apparent activation energy of biological systems varies with temperature, and seven references from the years 1926-1931 to show that these variations were systematic. He confesses (Bělehrádek, 1926a) to trying various mathematical formulae to systematise the influence of temperature on biological processes and found a simple power function the most effective general formula. This approach would be heartily endorsed by present day mathematical modellers (Ratkowsky, 1990) and had Bělehrádek kept strictly to mathematics his views might have been generally accepted. However, he tried to relate the powers obtained to such properties as protoplasmic viscosity (Bělehrádek, 1926b). The quotations in Table P.3 show the polarisation of thought vis à vis the Arrhenius and Bělehrádek approaches.

In microbiology, this dichotomy of thought continues up to the present day, although it must be appreciated that the Bělehrádek power functions have been used by zoologists, botanists and oceanographers for the last fifty years. An introduction to this literature is given in Ross (1987).

The Arrhenius equation has become sufficiently enshrined in microbiology to be totally accepted in the 1990 text "Physiology of the bacterial cell" by Neidhardt, Ingraham and Schaechter. These authors arbitrarily divide the Arrhenius plot of bacterial growth rates below the optimum temperature into a normal and minimum growth temperature portion to obtain Arrhenius kinetics in the *normal* region. At this point, it seems appropriate to take a quotation from the chapter "Equations for Everything" in Ian Stewart's most readable book "Does God play dice. The new mathematics of chaos" (1990). He quotes the article from the *Proceedings of the Royal Institution* 1986 by the British mathematician Christopher Zeeman relating to the discovery of an ancient Greek planetary calculator.

"First came the astronomers observing the motions of the heavenly bodies and collecting data. Secondly came the mathematicians inventing mathematical notation to describe the motions and fit the data. Thirdly came the technicians making mechanical models to simulate those mathematical constructions. Fourthly came generations of students who learned their astronomy from these machines. Fifthly came scientists whose imagination had been so blinkered by generations of such learning that they actually believed that this was how the heavens worked. Sixthly came the authorities who insisted upon the received dogma. And so the human race was fooled into accepting the Ptolemaic System for a thousand years."

The time span of current dogmas is one hundred rather than one thousand years, but the lessons are the same. Stewart points out that the

Table P.3 **Differences of Opinion**

"The biological problem of temperature coefficients suffered from the beginning from the fatal circumstance that it was not initiated by biologists, but by chemists. It has, of course, long been known both to plant and to animal physiologists that biological processes are accelerated by raising the temperature up to the optimum, but for a long time no attempt was made by biologists to study this common phenomenon on a more quantitative basis and to draw general conclusions as to its nature and meaning.

It was undoubtedly to the merit of some chemists, chiefly Van't Hoff, Cohen, Abegg, Herzog, Arrhenius, that they drew the attention of biologists to the possibility of a more quantitative and deeper study of the action of temperature on living organisms. At the same time, however, this purely biological problem has been misled by a chemical line of thought into a blind alley."

(Bělehrádek , 1930)

"One cannot expect that a chemical law, which does not hold good in enzymatic processes, should yield results in processes taking place inside the living cells. It may be said that the Van't Hoff-Arrhenius law has no real validity in biology."

(Bělehrádek , 1930)

"In contrast to those who have sought a physical-chemical interpretation of biological rate processes, a considerable number of investigators have concluded that the usual laws governing the temperature relations of chemical reactions do not hold in biology (cf. discussions in Bělehrádek , 1930). This conclusion smacks of vitalism and need not be considered in detail."*

(Johnson, Eyring and Polissar, 1954)

---

*The doctrine that biology needs to be explained in terms of the forces lying outside the domain of physics.

mid-20th Century concentrated on linear equations for the pragmatic reason that they couldn't solve anything else. Linearity, he says, is a trap.

The scientific approach seeks to solve complex problems by breaking them down and attempting to understand smaller components. In this way, the Arrhenius equation has become a deterministic model for bacterial growth. It should be remembered that the model originated as an empirical relationship for simple reactions and was subsequently given a theoretical basis by Eyring's Transition State Theory (Eyring, 1935). However, the simple Arrhenius model clearly falls down when applied to complex reactions, particularly biological reactions catalysed by enzymes, the catalytic efficiency of which is known to change with temperature. In this regard master-reaction models, which take into account the effect of temperature upon the catalytic efficiency of the enzyme itself, may provide a firmer theoretical basis and better description of the observed growth rate response to temperature.

As stated earlier, Bělehrádek had taken a more objective approach than the Arrhenius school of thought and had noted seven references between 1926 and 1931 referring to a *systematic* variation in rates of biological processes with temperature.

We hope that it will become apparent to the readers of this book that power functions as used by Bělehrádek are particularly appropriate for the description of the temperature dependence of the rate of biological processes. We also feel that the Bělehrádek and master reaction approaches are not necessarily contradictory, and that a common ground may exist between them.

Predictions of mean rates sometimes do not appear to translate well to predictions of time, such as shelf life at temperatures far from the optimum growth rate of bacteria. This is because the natural variance in time itself increases as a power and consequently an individual observation may well be away from the predicted mean. We now know that for many microbial growth processes the transformation of the equivalent root of the power may provide a convenient method for obtaining linear relationships with homogeneous variance, whilst in others a logarithmic transformation may be more appropriate to make the variance homogeneous. The reader must approach modelling with an open mind, with a clear goal and be aware that the use of logarithms and power functions will lead to predictable but different confidence limits for modelled values of rate or time for the same data.

Bělehrádek's original belief that there are *systematic* changes in the rate of life processes with temperature, however, opens up a new physiological perspective derived from some rather simple mathematics.

1.

INTRODUCTION

The concept of predictive microbiology is that a detailed knowledge of the growth responses of microorganisms to environmental conditions enables objective evaluation of the effect of processing, distribution and storage operations on the microbiological safety and quality of foods. If the microbial ecology of a processing operation or a product during distribution or storage is understood, survival and/or growth of an organism of concern may be predicted on the basis of a mathematical relationship between microbial growth rate and environmental conditions. In turn, this means that the potential exists to estimate food quality and safety by monitoring storage conditions such as temperature or gaseous atmosphere or intrinsic properties of the food such as water activity or pH value.

1.1. PREDICTIVE MICROBIOLOGY

The growth of spoilage or pathogenic microorganisms on foods is of concern to everyone involved in the production chain, from primary producers, to processors, distributors, retailers and ultimately the consumer. Whether from a health, profit, or other point of view, we can and must take precautions to restrict the growth of microorganisms, but it is still important to be able to assess the remaining shelf life or microbiological safety of a particular product at a specified stage in the chain. This may be to optimise food inventory management, to make decisions regarding the suitability of a particular product for a certain use, to evaluate processes, or to resolve disputes over product quality e.g. as a result of mishandling of product. Where products are being transported over long distances or stored for long periods these issues are especially important (Ross and McMeekin, 1991).

In the 1980s, the marked increase in the incidence of major food poisoning outbreaks led to an acutely increased public awareness of the requirement for a safe and wholesome food supply. During the same period many food microbiologists were beginning to accept that traditional microbiological methods to determine food quality and safety were limited by the time needed to obtain results and that indirect methods relying on chemical, physical or physico-chemical changes did not give a response until large numbers of cells were present. In the case of spoilage organisms, detectable responses occurring at ~10^7 cells/g or mL of product meant that the food was at the stage of incipient spoilage requiring only a few generations for overt spoilage to be evident. Similarly, many proposed rapid methods require growth to high levels before a response is evident, whilst others depend on the use of sophisticated and expensive equipment.

An alternative to developing technology to provide faster microbiological analyses, and one which may ultimately prove less costly, is predictive microbiology. Predictive microbiology relies upon the development of mathematical models which can predict the *rate* of growth or decline of microorganisms under a given set of environmental conditions. The models are constructed by measuring the responses of microorganisms to the main controlling factors such as temperature, pH, gaseous atmosphere, chemical preservatives and water activity, both singly and in combination. The models can then be used to predict the extent of growth or decline of the microorganism under conditions that were not specifically tested for in the original testing protocol. Reviews of predictive microbiology include Campbell-Platt (1985); Farber (1986); McMeekin and Olley (1986); Baird-Parker and Kilsby (1987), and Gould (1989).

The possibility of predicting microbial behaviour in foods is not new and reference to the use of predictive microbiology can be found in the literature of the 1930s. Scott (1937) reported on the effect of temperature on the growth of microorganisms on ox muscle. The introduction to his paper, reproduced below, demonstrates clearly that he understood the potential to use accumulated kinetic data on microbial growth responses to predict the shelf life and safety of foods.

"A knowledge of the rates of growth of certain micro-organisms at different temperatures is essential to studies of the spoilage of chilled beef. Having these data it should be possible to predict the relative influence on spoilage exerted by the various organisms at each storage temperature. Further, it would be feasible to predict the possible extent of the changes in populations which various organisms may undergo during the initial cooling of sides of beef in the meatworks when the meat surfaces are frequently at temperatures very favourable to microbial proliferation."

Considerable international interest in predictive microbiology has developed, particularly in the United States of America and the United Kingdom, and also Australia and Western Europe. Governments, research organisations and industry are working both independently and collaboratively on programs to generate the databases required for the development of predictive models, and also on systems for making that information accessible in a readily usable form to a wide range of users. Cooperation between nations and hitherto competing institutions is occurring.

In the UK the Ministry of Agriculture, Fisheries and Food (MAFF) initiated a nationally coordinated program of research into the growth and survival of microorganisms in food systems. Gould (1989) reported that the aim was "to build on the new modelling expertise that is becoming available and to generate a computerised Predictive Microbiological Data Base that will be ahead of any similar development in the rest of the world." The project, anticipated to cost ~£9 million over a six year period, involves the AFRC Institute of Food Research Laboratories, Campden Food and Drink, Leatherhead Food and the Flour Milling and Baking Research Associations, Torry Research Station, Unilever Research, the University of Bath and the University of Wales, Cardiff. The project targets food-borne pathogens: *Aeromonas, Bacillus cereus, Campylobacter* spp., *Clostridium botulinum, Escherichia coli* 0157:H7, *Listeria monocytogenes, Salmonella* spp., *Staphylococcus aureus* and *Yersinia enterocolitica.* Growth, death and survival responses as affected by temperature, pH, organic acid, NaCl, water activity, preservatives and gaseous atmosphere are under consideration. The collected data are collated in a single database at Campden Food Research Association where they are analysed by a "models base" to provide information for an expert system which can be accessed directly by non-expert users. Further details of the program are given by Walker and Jones (1992).

Interest in Europe is being developed through the FLAIR (Food Linked Agricultural and Industrial Research) program with ~30 laboratories in 10 EEC countries acting collaboratively to examine growth responses of spoilage and pathogenic organisms in a wide range of natural products.

In the United States of America predictive microbiology research is centred on the USDA in Pennsylvania where the Microbial Food Safety Research Unit has described the effect of five variables (temperature, NaCl concentration, pH, nitrite concentration and gaseous atmosphere) on the growth of the food-borne pathogens *Listeria monocytogenes, Salmonella, Clostridium botulinum, Aeromonas hydrophila, Shigella flexneri*, *E. coli* 0157, and *Yersinia enterocolitica*. The databases held by this group are the product of an initial five year programme and provide a basis for continuing modelling studies. The *L. monocytogenes* database alone is estimated to have cost ~US$500 000 (R.L. Buchanan, *pers. comm.*). This level of research expenditure is quickly put into perspective by the enormous costs of food poisoning outbreaks. The costs accrue due to investigation and testing, health care, costs associated with loss of productive activity, compensation for illness (and death), and costs to the industry due to lost production and loss of consumer and market confidence. Socket (1991), in an analysis of these factors for *Salmonella* infection in the UK in 1988, estimated the cost at £18 144 000 made up of public sector costs (£6 849 000), costs to families (£1 796 000), and lost production (£9 499 000).

The potential for savings through prevention of spoilage is also enormous. The losses are difficult to quantify, but in Australia estimates range up to 20%. This is significant in an industry which is that country's largest manufacturing activity, accounting for 20% of total manufacturing sales, employing 170 000 people in 3 600 firms and having an annual turnover of A$24 billion (Department of Industry, Technology and Commerce, 1989).

Predictive microbiology should not be considered to be relevant only to the food industry. Biotechnologists seeking to optimise fermentation processes will benefit from being able to model the differential effects of environmental factors on yields from particular processes. Products of a variety of microbial processes such as algal biomass, or specialised enzymes able to withstand the rigours of industrial processes, are likely to benefit from the advance of predictive microbiology; so too other processes such as mineral leaching, bioremediation and microbial treatment of effluents. The quantitative approach to microbial ecology afforded by predictive modelling might also be applied to understanding and minimising the deleterious effects of microbial rich effluents, biofilms, microbial blooms, etc. Models describing survival and death of microorganisms will also be of relevance in the pharmaceutical industry.

1.2. THE MODELLING PROCESS

An absolute prerequisite to the development of reliable mathematical models is the collection and appropriate processing of good quality data describing the effects of different factors on microbial development. For the full benefits of predictive microbiology to be attained appropriate technology will also need to be developed. This will almost certainly take the form of interactive computer software based on mathematical models summarising large databases of growth responses, and may involve the development of smart sensors which record and interpret the effect of environmental parameters on the growth of microorganisms of interest in a product or process. Thus the successful implementation of predictive microbiology requires skills in microbiology, mathematics and statistics, information science, chemistry and engineering. There is scepticism on the part of many potential users of predictive microbiology that predictive models which do not have a basis in fundamental biochemical, physiological and thermodynamic theory, are intrinsically unreliable. Whilst we would strongly argue against this, we acknowledge that there is also the need for skills in these areas to be applied to the study of predictive microbiology. This book itself is the result of collaboration between microbiologists, a chemist and a statistician. If there is one theme that we would wish to emphasise as a result of our often traumatic interaction, it is that a multidisciplinary approach, however frustrating, will be required for the successful implementation of the predictive microbiology concept.

The need for all these skills to be applied concurrently is another lesson we have learnt. Dr. John Stocker, the chief executive of CSIRO Australia, has exemplified this in his strong plea for industry to be involved in the early stages of scientific advance. It is too late to take a concept to industry when, for example, a temperature function monitoring strip with the wrong temperature characteristic is already on the production line. The much used phrase, 'technology transfer' is perhaps inappropriate - we should be integrating the technological needs of research into the earliest phases of research planning.

1.3. PREDICTIVE MICROBIOLOGY AND HACCP

Proponents of predictive food microbiology are encouraged by the acceptance of HACCP (Hazard Analysis and Critical Control Points) systems to control contamination in food processing operations, and by the general view in the food industry that HACCP represents a superior approach to that offered by traditional inspection and end-product testing for quality and safety assurance of food. HACCP and predictive microbiology are complementary concepts and predictive microbiology can enable a more powerful use of HACCP by permitting objective and quantitative assessment of the effect of processing steps on product quality and safety.

ICMSF (1988) defined the HACCP system in terms of:

1. Hazard analysis (HA) - identification of hazards and assessment of their severity at various stages of food procurement, processing, distribution, storage and end use.
2. Critical control points (CCP) - points at which control can be exercised to prevent a hazard (CCP1) or minimise the risk of a hazard (CCP2).
3. Specification of criteria such as intrinsic, extrinsic or processing factors to ensure control or biological criteria to indicate that control has been achieved.
4. Establishment and implementation of procedures to monitor control points and to respond rapidly to non-compliance situations.

The sequence of events listed are equivalent to those proposed in predictive food microbiology studies (McMeekin *et al.*, 1992):

1. Identify the organism of concern (spoiler or pathogen).
2. Develop an understanding of the ecology of the organism to enable identification of the sources and likely level of contamination and the effect of processing, distribution, storage, etc. on numbers.
3. Compare the information obtained with preset criteria to accept/reject the process or product or to indicate the remaining shelf life.
4. Incorporate the information into monitoring devices that indicate the extent of microbial proliferation.

Thus predictive food microbiology can be viewed as an extension of the HACCP concept in which a kinetic model of the organism's response to its ecosystem is interpreted by sensors responding in the same way as the organism to changing environmental conditions. The sensor reading provides a direct measure of the stage and extent of microbial growth and specifications can be set from zero growth (lag phase not resolved) through any number of generations at which a public health or spoilage risk is perceived.

1.4. ABOUT THIS MONOGRAPH

Because of the quantity of data required, modelling studies usually entail observations made under well-controlled conditions using laboratory media. Growth rates or lag phase duration may be determined by a variety of methods. The data are usually fitted to a sigmoid growth function from which estimates of generation time, lag phase duration or time to a specified increase in bacterial numbers are derived. These values are then used to generate a model describing the effect of environmental factors such as temperature, water activity, pH, etc. on the rate of microbial development. In many studies, kinetic models, such as Bělehrádek-type models, Arrhenius-type models or polynomial models are used. It is this aspect of predictive microbiology with which this text is concerned. Alternatively, the probability of microbial growth or toxin production under a defined set of conditions may be predicted.

When a basic model has been derived as a result of laboratory studies, it is essential to validate the ability of the model to describe accurately microbial development in food under the expected conditions of storage. Validation trials should include fluctuating conditions, particularly of temperature, which is the situation normally encountered in the distribution and storage of most foods.

To apply the validated model in practice, the information contained in the model needs to be incorporated into a monitoring device that mimics microbial response to changing environmental conditions. Several types of monitors have been used, including chemical and physical indicators, and electronic devices that display the cumulative temperature history or that allow the information to be collated and assessed by computer software packages.

The major steps in model building, validation and application, provide a structural basis for this text.

The basic concepts and practical considerations involved in model building are dealt with in detail in Chapter 2. In retrospect, it was inevitable that this would become the longest chapter in the monograph. Its development is largely attributed to one of the authors (T. Ross) who was required to examine modelling and experimental protocols as an introduction to a PhD program in predictive microbiology. In doing so, attention was focussed on the various stages of model building and the effect that often overlooked experimental details, such as preparation of inocula, might have on the data collected for subsequent analysis.

A variety of methods of enumeration and analysis of growth rate data are considered with an examination of the way in which experimental data is fitted to the model type chosen. Sections on validation and comparison of models precede guidelines for nonlinear regression modelling prepared by D. Ratkowsky.

In Chapter 3 readers are introduced to the two major types of temperature dependent kinetic models (the Arrhenius-type and the Bělehrádek-type). A comparison of the ability of both types to handle real data is provided in Chapter 4 with attention to the guidelines outlined for model comparison in Chapter 2.

To model the combined effects of temperature and other constraints on microbial growth, extensions of Arrhenius- and Bělehrádek -type models and polynomial models are considered in Chapter 5.

Validation and application studies are the subject of Chapters 6 and 7 which detail attempts to use predictive microbiology to assess the remaining shelf life of chill stored foods and the evaluation of hygiene during meat processing operations.

Devices for practical application are dealt with in Chapter 8 (time-temperature indicators based on physical characteristics such as the diffusion of waxes or chemical reactions) and Chapter 9 (electronic monitoring devices). In relation to the latter, basic concepts of mathematical integration, electrical transducers and electronic memory are considered with a description of commercially available devices and predictive microbiology software.

Driven by our perception that empirical predictive models continue to be viewed with mistrust, the monograph concludes (Chapter 10) with a thermodynamic treatment of bacterial growth and the consequences of such a treatment for the kinetic models in contemporary use.

1.5. AN UNDERLYING CONCEPT?

Predictive microbiology as a discipline has developed through experiments to measure the change in growth rate of organisms in response to temperature. Little work has been done specifically to develop equations for the production of metabolites by bacterial or intrinsic enzymes in foods. It is interesting to note that the original relative rate curve for spoilage of proteinaceous foods was compiled by two of the present authors (Olley and Ratkowsky, 1973a and b) from 35 data sets involving chemical metabolites, 15 data sets on bacterial counts, 15 data sets on organoleptic assessment of chemical change (visual, olfactory and gustatory) and five data sets involving gross physical changes in products.

A complex mixture of intrinsic and bacterial changes in nucleotides was studied by Japanese workers (Ohta and Hirahara, 1977). These data could be linearised by the square root equation, as could the original universal spoilage data up to 15°C and the response of psychrotrophic bacteria to temperature (Ratkowsky *et al.*, 1982).

Bremner *et al.* (1987) also alluded to the universality of the temperature-rate relationship: “There is a strong case that at least two of the major determinants of deterioration in chilled seafoods, namely microbial spoilage and nucleotide breakdown, follow similar temperature functions. What evidence there is for other phenomena indicates that the sum total of all the degradative processes, to all intents and purposes, also follows a similar pattern up to 15°C when other factors such as spoilage due to mesophiles may intrude".

The apparent universality of the rate response of 'natural' process to temperature suggests some underlying common phenomena, and also that research into microbial growth rate responses to temperature (and other factors) may have wider applicability. For this reason, and others discussed above, a text such as this should hold interest for a broader spectrum of readers than just food microbiologists.

1.6. SUMMARY

We believe that this monograph provides a contemporary view of the status of kinetic predictive microbiology, both from a theoretical and an applied perspective. The monograph is both a review and a practical manual. Much of what is presented will be also be applicable to other types of predictive modelling of microbial responses.

Work on the book started in early 1989 and since then there has been an explosion of interest in the subject leading to the publication of a variety of predictive models. As a result of this dynamic situation the conscientious reader may perceive a number of conflicting philosophical strands which have been difficult to reconcile with individual authors' points of view. A feeling of searching may be found in some chapters, particularly the last few pages of Chapter 10 which were written in October 1992 at the end of a four year gestation period.

Nevertheless, the monograph reflects the authors' training and backgrounds, including mathematical, microbiological and food technology perspectives. We hope to impart to the reader our belief that predictive microbiology is, like all good ideas, quite simple in theory. In its application, however, there are a number of 'traps for young players'; through our experience others may avoid these pitfalls and the successful adoption of predictive microbiology be made easier. Worldwide interest in the subject, major research programs in many countries and increasing international collaboration to ensure a safe and wholesome food supply lead us to the conclusion that predictive microbiology will become increasingly accepted as a useful and practical concept by the food industry, regulatory authorities and other interested parties.

2.

BASIC CONCEPTS AND METHODS

Means of collecting and analysing growth rate data are briefly explained and critically reviewed from the perspective of collecting reproducible and comparable kinetic data. Philosophies for the development and validation of reliable models, and criteria for comparison and selection of competing models are also presented and reviewed.

2.1. INTRODUCTION

Causton (1987) states that the essential purpose of mathematical models is to describe succinctly a set of acquired data. It is more correct, however, to consider the model as describing an underlying process which generates data. Once constructed, the model can be used for many purposes which are extensively discussed in subsequent chapters in relation to predictive food microbiology.

Two major types of model are recognised. **Empirical** models are derived from an essentially pragmatic perspective. They simply describe the data in a convenient mathematical relationship and consequently often give little or no insight into the underlying process. **Mechanistic,** or **deterministic,** models are built up from theoretical bases and, if they are correctly formulated, may allow the interpretation of the modelled response in terms of known phenomena and processes.

2.1.1. The Modelling Process

The various stages of model development have been identified by Draper and Smith (1981), and summarised by Farber (1986) as:

i) Planning
ii) Data collection and analysis
iii) Mathematical description
and iv) Validation and maintenance

and are the same stages less explicitly discussed by Baird-Parker and Kilsby (1987), Roberts (1990), and Ross and McMeekin (1991).

From the literature it is apparent that there exists a spectrum of strategies for developing predictive models for food microbiology. There are different types of problem (toxin formation, shelf life/spoilage, pathogen growth, death kinetics); models of different types (e.g. probabilistic, kinetic); various methods for collecting data (optical density, viable counts, conductimetry, metabolite assays); several mathematical expressions used to describe what is essentially the same type of data (see Chapters 3, 4 and 5); and there have been a number of indices suggested for evaluating models (Adair *et al.*, 1989; Bratchell *et al.*, 1990; Zwietering *et al.*, 1990) some of which have attracted comment by various authors (McMeekin *et al.*, 1989; Davey, 1989a; Kilsby, 1989; Zwietering *et al.*, 1991; Heitzer *et al.*, 1991).

This chapter will be concerned primarily with kinetic models although some of the discussion will also be relevant generally to other model types. We will present basic concepts concerning growth kinetics of bacterial populations and ways of describing them, review different strategies, methods, and technology for collecting growth data from which to derive kinetic models, introduce modelling methods, and suggest guidelines for modelling and criteria for the evaluation of competing kinetic models.

2.1.2. Terminology

Throughout this text we will of necessity use mathematical terms with which some readers may be unfamiliar. Several of the more fundamental and frequently used terms are explained below.

2.1.2.1. Variables, residuals, error terms

Independent, **explanatory** or **predictor** variables are those which it is believed will explain the type and magnitude of the response observed. In the case of kinetic models for food microbiology, these will typically be temperature, pH, water activity, and other agents in the modelled system which affect the rate of growth of the modelled organism. **Response** or **dependent** variables are those properties of the system which are governed by the independent variables. In the case of kinetic models, this will usually be the rate or duration of some microbial growth process.

A model is formulated to describe the known or assumed qualitative relationship between the predictor and response variables, but it is the parameters of the equation which quantify the relationship. Parameters are constants for a given set of experimental conditions but must be estimated from the data to give the best possible fit to the data. The usual means of estimation of parameters is the minimisation of the values of the **residuals**, which are the differences between the observed values of the response variable and those predicted by the fitted equation.

There is always some error inherent in any measurement of an observation. This error is the difference between the particular observation and the mean, or predicted, response. Accordingly, in any mathematical description of a data set there will also be some error between the predicted and observed values. The behaviour of the error can be investigated and described. For example it may, on average, be a constant value irrespective of the magnitude of the response. Alternatively it may be a function of the magnitude of the response. Description of this behaviour is an integral part of the modelling process and is expressed in the **error**, or **stochastic,** term of the model. The **deterministic** part of the model describes the relationship between the variables. The significance of the stochastic assumption is discussed in 2.4.1.2. and 2.6.2.4. These features of the mathematical expression of models are shown in Fig. 2.1.

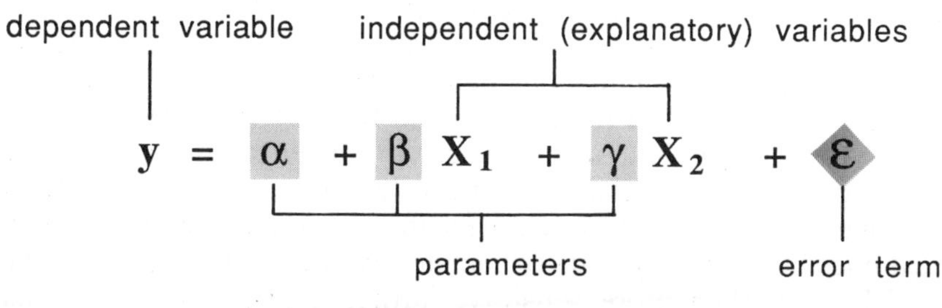

Fig. 2.1 An example of a mathematical model showing the nomenclature of the component terms. The values of the independent variables (X_1, X_2) are known or set before the response (y) is observed. The values of the parameters (α, β, γ) are determined by the data, and are fitted to minimise the difference, in some sense, between the observed response and that predicted by the model. The stochastic term (ε) indicates the extent to which the predicted response deviates from the observed response.

2.1.2.2. Linear and nonlinear

Readers should note that a 'curvilinear' model, such as

$$\mathbf{y} = \alpha + \beta\,\mathbf{X} + \gamma\,\mathbf{X}^2$$

is classified as a linear regression model, since the parameters α, β, and γ appear linearly, although the relationship between the response variable **y** and the explanatory variable **X** describes a curved line. In nonlinear models the parameters do not appear linearly. The important differences between these two types of model are the manner in which they are fitted to data, and their estimation properties once fitted. These topics are more fully explored in 2.4.

In building regression models to describe natural processes, it is likely that the appropriate model will prove to be a nonlinear one. This is because physical, chemical and biological models are often solutions of differential equations. Such solutions are rarely of the form that the model is linear in its parameters. Draper (1988) also states that mechanistic models are often nonlinear in their parameters and more difficult to fit and evaluate, and that appropriate experiments are more difficult to design. Some nonlinear regression models which have been proposed for use in predictive microbiology include the sigmoidally shaped Gompertz curve (Gibson *et al.,* 1988; Buchanan and Phillips, 1990), the square root models (Ratkowsky *et al.,* 1982, 1983), the Johnson and Lewin model (1946), the Sharpe and DeMichele (1977) model and its reparameterisation by Schoolfield *et al.* (1981).

Empirical models usually take the form of first or second degree polynomials. Polynomial models of higher order are rarely fitted partly because of the difficulty in interpreting the form of the fitted response surface (see 2.2.2.) which, in any case, produces predictions whose standard errors are higher than those from lower order models (Draper, 1988). Draper (1988) considers that mechanistic models are preferable to empirical ones because they usually contain fewer parameters, fit the data better, and extrapolate more sensibly, further noting that polynomial models often extrapolate poorly. We will return to these considerations later in this chapter.

2.2. PLANNING

2.2.1. Introduction

"The specific statement of the problem is the most important phase of any problem solving procedure (Draper and Smith, 1981)." Given the effort required to generate a model of even moderate complexity this is sound advice. It may be possible to simplify the amount of experimental work which needs to be done simply by thinking about the problem or doing some preliminary experiment(s) beforehand. For example, it may be thought desirable to generate a model to evaluate the extent of growth of a pathogen in a particular product. If the normal spoilage organisms present grow more quickly than the pathogen under all conceivable storage conditions so that the product always spoils before the pathogen reaches infectious dose levels, or, if the spoilage biota produces a metabolite completely inhibitory to the pathogen of concern and completely prevents its growth in that product, there would be no point in constructing a model. Similarly, for foods which are produced to a uniform standard of quality, the direct modelling of microbial growth in the product may be obviated by recourse to the relative rate concept discussed in 6.2.

The nature of the organism of concern and the problem it causes will partly determine the type of model required. Different types of model are used to predict the probability of toxin formation, the probability of growth, the duration of lag time of a pathogen for which there is zero tolerance, the extent of growth of a pathogen for which some tolerance is allowed, and the extent of spoilage or proliferation of spoilage organisms. In turn, each of these model types has particular experimental strategies appropriate to its development.

Factors that need to be considered when designing experiments to gather the data upon which the model is based are the variables to be included; whether there are interactions between variables; the range of values of the variables; the amount of data required; the experimental model used and the response to measure and model.

2.2.2. Independent Variables

When developing models to describe the growth rate response as a function of several independent variables, we are modelling a 'response surface' in n dimensions where n is the number of independent variables. A response surface design is simply an arrangement of experimental points representing possible combinations of variables in that n~dimensional space. The choice of response surface design is thus one of selecting a set of suitable combinations according to some preselected criterion of goodness (Draper, 1988). Desirable features of such a design are given in Table 2.1, opposite.

2.2.2.1.Interactions

If it is necessary to model the effect of a number of variables acting simultaneously, it may be useful to do a factorial experiment first to determine whether there are any interactions among the factors. Typically this is done by having each variable either present at a single level or absent, and testing the response at each combination of the variables, that is, using 2^n test conditions where ‘n’ is the number of variables. Factorial experiments may also be designed to include many values of the independent variables. The analysis of factorial experiments is fairly straightforward and is discussed in texts on statistics or experimental design (for example Cochran and Cox, 1957).

2.2.2.2.Range of values

The range of the variables to be included in the experimental design will depend upon the range expected to be encountered by the product. It is generally accepted that it is unwise to extrapolate beyond the range of the data used to generate the model so that the full range expected must be tested. It is also sound practice to use as full a range of the independent variable as possible to obtain as full a range of the response as possible. Even in situations where the expected range of values of the independent variable is small, using a wider range may reveal behaviour that could not be resolved over a narrower range.

Statistical principles should also be considered when determining the number of levels of independent variables to be tested. When many independent variables are being evaluated the number of test conditions rapidly multiplies, so that decisions have to be made regarding the feasibility of conducting all experiments concurrently, which must also be considered in relation to the method used to measure the response. Various strategies have been proposed to deal with this problem. Ross and McMeekin (1991) propose methods to reduce the workload involved in collecting data upon which to base the model so that the entire response surface can be tested. Various workers have used a central composite design to reduce the number of experimental points that are tested on the response surface (Gibson *et al.*, 1988; Buchanan and Phillips, 1990; Palumbo *et al.*, 1991). The central composite design is

Table 2.1 **Desirable Features of Response Surface Designs**

The response surface design should:

1. generate a satisfactory distribution of information about the behaviour of the response variable throughout a region of interest, R
2. ensure that the fitted value at **x**, $\hat{y}$ (**x**), be as close as possible to the true value at **x**, η(**x**).
3. give good detectability of lack of fit
4. allow transformations to be estimated
5. allow experiments to be performed in blocks
6. allow designs of increasing order to be built up sequentially
7. provide an internal estimate of error
8. be insensitive to wild observations and to violation of the usual normal theory assumptions
9. require a minimum number of experimental points
10. provide simple data patterns that allow ready visual appreciation
11. ensure simplicity of calculation
12. behave well when errors occur in the settings of the predictor variables
13. not require an impractically large number of predictor variable levels
14. provide a check on the constancy of variance assumption

[*After* Draper (1988)].

described by Draper (1988) who comments that it has many of the desirable features referred to in Table 2.1. For some applications, however, one may be more interested in the response at conditions far removed from the centre of the range. For example, many food processors will be more interested in the responses of microorganisms at near growth-limiting conditions, as these are the conditions which will extend the shelf life and safety of their products. In this situation most observations are required near the extremes of the range, where growth is retarded, rather than the centre of the range.

2.2.3. Preparation of Inocula

Various published reports on the growth responses of microorganisms have drawn attention to the fact that the condition of the inoculum may have an effect on the subsequent growth kinetics, particularly the lag phase duration. In this section we will discuss some of the potential problems, possible interpretations, and methods for controlling variable features of the inoculum itself.

2.2.3.1. Physiological status

Kinetic models attempt to explain the time taken for a specified growth response (see 2.2.4.) in terms of environmental variables such as temperature, water activity, pH, etc. Recent reports (Walker *et al.*, 1990; Buchanan and Klawitter, 1991) have suggested that growth responses, particularly lag time duration, may also be dependent upon the prior temperature history of the inoculum used for kinetic experiments. Fu *et al.* (1991) claim to have demonstrated temperature history effects on the generation time of *Pseudomonas fragi*. Other reports (Shaw *et al.*, 1971; Charm *et al.*, 1972; Pooni and Mead, 1984; Gill and Harrison, 1985; Gill, 1986; Blankenship *et al.*, 1988) suggest, however, that exponential growth rates are independent of temperature history.

We believe that an explanation may lie in the physiological condition of the inoculum used, but until this issue is resolved it is essential, when comparing results of kinetic experiments, to ensure that the inocula used were physiologically comparable, particularly when modelling the lag phase of growth. Temperature history effects on lag times may be a consequence of inocula, grown at different temperatures, which were not physiologically identical because the time-temperature combinations were not equivalent. This is illustrated below. The calculation of equivalent time-temperature regimes is discussed in 9.3.2., but requires temperature-rate data for the organism of interest.

One means to avoid the problem is to perform all experiments simultaneously using a single culture for the inoculum. In this situation it may still be necessary to take precautions such as reducing or stopping the growth of the inoculum, (for example by reducing the temperature to that at which the growth rate is negligible), prior to and during the inoculation procedure so that the inoculum density does not increase during the time taken to inoculate all cultures. In many cases, however, it is not possible to perform all determinations simultaneously and special precautions must be taken to prepare standardised inocula.

The use of exponential phase cells may simplify the problem also, but caution must still be exercised. For example, a common expression in the literature is that of 'an overnight culture' which is intended to indicate a culture in the exponential phase of growth and which is, by inference, physiologically homogeneous. Consider, however, a typical mesophilic enteropathogen with a generation time at 37°C of, say, 30 min. Assume that 'overnight' is about 14 h (i.e. from 6:00 p.m. one day to 8:00 a.m. the following morning), that a typical inoculum is 10^6 cells into 100 ml laboratory media, and that the culture achieves a maximum density (stationary phase) of 10^9 cells/ml. Given these circumstances the culture would have reached the stationary phase after about 8 - 8.5 h. Even at 25°C, or if the inoculum were only 10^3 cells/ml, the culture may be expected to have reached stationary phase within the 'overnight' incubation. Clearly care must be exercised to ensure that the cells are in the condition or phase intended, otherwise the inoculum may represent an uncontrolled variable. Further examples below illustrate that considerable attention must be paid to this detail.

Walker *et al.* (1990) investigated a temperature history effect on the lag time of *Listeria monocytogenes* strains. The preincubation conditions were reported by these authors, to result in "approximately the same level of inoculum and phase (mid-late logarithmic) of growth". Analysis of the growth rate data presented by them enables the construction of a temperature-rate model ($R^2 = 0.99$) of the square root type (Eqn. 3.10) which predicts a generation time at 4°C of 22 h and which, by extrapolation (see 2.2.2.2), predicts a generation time at 30°C of 1.1 h. Though potentially unreliable, it is unlikely that the extrapolated generation time is too short: Buchanan *et al.* (1989) report generation times for *Listeria monocytogenes* Scott A at 28°C and 37°C of 0.5 - 0.6 h under similar growth conditions. Using the data of Walker *et al.* (1990), the simple calculations below suggest that the 30°C incubation is, on a relative basis, 6.7 times as long as that at 4°C.

4°C: Generations = time / generation time

$$= (6 \text{ days} \times 24 \text{ h.day}^{-1}) / 22 \text{ h.gen}^{-1}$$

$$= 144 \text{ h} / 22 \text{ h.gen}^{-1}$$

$$= 6.5 \text{ generations}$$

30°C: Generations = time / generation time
= 48 h / 1.1 $h.gen^{-1}$
= 43.6 generations

Comparison: 43.6 / 6.5
= 6.7

Buchanan and Klawitter (1991) used various time-temperature combinations to prepare inocula of *Listeria monocytogenes* Scott A prior to incubation at 5°C, in a study of temperature history effects. Again, the data of Buchanan *et al.* (1989) for the growth of this strain detail observed generation times corresponding to some of these temperatures and, by interpolation between these values in accordance with a square root model (see Eqn. 3.10), allow the estimation of the generation times at the other temperatures. Table 2.2 indicates the calculated number of generations for each pre-incubation condition. Forty or more generations is probably not possible in a typical batch culture, but the results of Table 2.2 serve to indicate that not all inocula had experienced equivalent time-temperature conditions.

Table 2.2 **Number of Generations of *Listeria monocytogenes* Expected Under the Inoculum Incubation Conditions of Buchanan and Klawitter (1991).**

Temperature (°C)	Time (h)	Predicted Generations
5	144	10
10	96	26
13	96	42
19	48	44
28	24	44
37	24	40

In some situations it will be desirable to use inocula in lag phase or from stationary phase. In these cases the steps needed to ensure inocula of an equal number of physiologically comparable cells would be to incubate a specified number of exponential phase cells at the required temperature for a time appropriate to the growth rate at that temperature, so that an equivalent number of generations occurred. Exponential phase cells are used as the initial inoculum as they represent an 'easily' reproducible physiological state, provided that the same growth medium is used. Inocula prepared in this manner may also form the basis of controlled 'temperature history effect' experiments.

The above discussion is not intended as a critique of reported temperature history effects studies but is included to emphasise that it is necessary to plan carefully the preparation of inocula, and to illustrate a mechanism whereby equivalent time-temperature conditions may be determined.

2.2.3.2. Inoculum size

The possibility that inoculum size has an effect on the growth responses of a culture such as lag time, exponential growth rate or maximum population density has also been addressed. In general there appears to be no effect of inoculum size on exponential growth rate or maximum population density (Jason, 1983; Buchanan and Phillips, 1990; Ross, unpublished). One important point to note is that it is essential to distinguish between the detection time of a response and the lag time. All practical bacterial enumeration methods have a lower limit of detection. If the inoculum density used is below this level then the time taken for an observable change in bacterial numbers is a combination of the lag time and the time taken for the population size to increase to a detectable level.

2.2.3.3. Mixtures of strains and species

Some workers (Gibson *et al.*, 1987; Gibson *et al.*, 1988) prefer to use a mixture of strains of the modelled organism, arguing that it is more representative of the situation that is likely to pertain to contamination of real foods. Such mixtures are sometimes referred to as 'cocktails'.

Consideration of the dynamic properties of mixed populations (see Appendix 2A.1) suggests that a faster growing strain/organism will soon become the dominant component of the mixed population, and that the observed growth rate for the population will, as the sub-population of that strain increases, rise to that of the fastest growing organism.

The results of Hayward (1990) lend support to this hypothesis. Superimposition of the square root plots for the axenic growth of six strains of *Aeromonas hydrophila* gave rise to a square root plot very similar to that for the growth of a culture containing all six strains initially in equal proportions.

Consequently we believe that a better alternative is to model several strains independently and to choose from those the strain that represents the worst case, i.e. that strain which grows fastest under the conditions of most interest. In the case referred to above the strain of *A. hydrophila* with the lowest T_{min} (see Eqn.3.10) was found to represent the worst case as it would grow at lower temperatures than the other strains, and had a faster growth rate at temperatures up to ~10°C. At 27°C (near to the optimum growth temperature), the growth rate predicted for this organism, strain 3459, was 15% slower than that for the cocktail. Such a temperature represents gross temperature abuse, however, and the difference in predicted generation times at such temperatures is likely to have no significance in practice, i.e. the worst case model should be 'fail safe' in practice.

2.2.3.4. Preparation of inocula for application to foods

When inoculating foods with organisms of interest to validate laboratory media based models, experimenters should be alert to the possibility of organisms growing in a surface film of the suspending media deposited on a food product. We suggest that only washed cell preparations should be used as inocula so that the observed growth rate is representative of the conditions experienced by the organism on the food, that steps be taken to minimise the amount of suspending medium applied to the food, and that the suspending medium should ideally mimic the pH, water activity, and preservative concentration of the food also.

2.2.4. Response Variable.

The primary response measured is usually the change in bacterial population density over time. The rate of change is then typically expressed as lag phase duration, generation time, or time to reach some specified level or condition, although the response is often modelled with rate (i.e. 1/response time) as the dependent variable. For reasons discussed below, the response variable is often mathematically transformed. For example, the response variable of the square root model is $\sqrt{rate}$, and of Arrhenius based models is usually ln (*rate*). Zwietering *et al.* (1991) presented a comparison of various proposed kinetic models, but, for stochastic reasons (see 2.4.1.2., 2.6.2.4.) expressed them all as forms with rate as the response variable.

In addition to the obvious response variables of lag and generation rates, other response variables which have been modelled include the square root of the reciprocal of the time taken for the culture to achieve a 30% change in optical density (Chandler and McMeekin, 1989a; Adams *et al.*, 1991); the logarithm of the reciprocal of the time taken for the culture to reach the most rapid rate of growth, given by the Gompertz M parameter (see 2.3.3.2.) (Gibson *et al.,* 1987; Buchanan *et al.*, 1989); various transformations of the time taken to achieve a specified increase in population numbers, for example; an increase by a specified number of $\log_{10}$ cycles (Stannard *et al.,* 1985; Gibbs and Williams, 1990), the time taken to reach a 'specified spoilage level' (see 6.1) indicated organoleptically or by a specified microbial load (Daud *et al.,* 1978; Gibbs and Williams, 1990), organoleptic scores (Bremner *et al.*, 1987), metabolite levels (Olley and Ratkowsky, 1973a) and time to reach a specified point on the modelled growth curve (Phillips and Griffiths, 1987). Lag phase duration and doubling time of a food spoilage yeast were modelled by Cole and Keenan (1987). Adair *et al.* (1989) used the natural logarithm of response time. For probabilistic models the response variable is the probability (as a function of the independent variables of storage conditions, product parameters and time) that some event such as growth initiation or formation of toxin will occur.

The form of the response variable chosen will depend upon the nature of the problem under investigation (growth rate, probability) and the model chosen to describe the response. For example, the use of logarithm of rate originally stemmed from the adoption of the Arrhenius model to describe bacterial growth responses. Prior to the development of computers and nonlinear regression software, the parameters of this model were best estimated from a graph of logarithm of rate against the reciprocal of absolute temperature. Similarly, for the square root model the fitting process is simplest (and better statistically, see 2.4.1.2.) when the data are expressed with square root of rate as the response variable as this leads to a 'straight-line' relationship between the response and explanatory variables.

With the advent of computer software which can perform complex fitting processes, one can be more objective about the selection of appropriate response variables. The more important issues relate to the stochastic properties of the selected response. These are more fully discussed later (see 2.4.1.2., 2.6., 4.4.). Stated succinctly, to obtain the best fit of the model to the data, the error in the estimate of the selected response must be independent of the value of the response or, if there are compelling reasons for the use of a response variable which does not have this property, the error behaviour must be compensated for by a weighting in the fitting process to give each datum point equal influence in the estimation of parameters.

2.2.5. Experimental Model

The experimental system used in the development of predictive kinetic models will always be based on a growth medium. In practice this will usually be a laboratory medium or some type of food. Determination of growth responses in liquid media allows a greater range of detection methods. In particular, indirect methods can be used which permit many more determinations to be made for a given amount of effort compared to direct enumeration methods. The organism will also most likely be more uniformly distributed than on a solid medium. A criticism of liquid laboratory media, however, is that they may not accurately model conditions in real foods, so that the model will subsequently have to be validated in the food product to which it is intended to be applied (see 2.5.).

Conversely, using real foods is both costly and logistically very difficult, and in general limits the method of measurement to viable (plate) counts or most probable number (MPN) methods. There are also difficulties in applying the test organism to the food in a consistent and representative manner. Furthermore, all available test methods are invasive. Accordingly, unless a liquid food is used, an extra element of variability is introduced from the growth medium as each determination must be undertaken on a unique sample of that food. Food-based experimental models may also lead to complications due to the presence of normal spoilage biota, and may generate models specific for the particular food substrate used.

Some of these problems were encountered and discussed by Gibson *et al.* (1987, 1988) particularly in relation to obtaining sufficient reliable data to model the growth curve so that objective measurements of growth parameters could be derived (see also 2.3.2.4.). For the above reasons Ross and McMeekin (1991) concluded that kinetic models were best developed by measuring response rates at close intervals of the independent variables in laboratory liquid media. The model may then be validated by comparison of data obtained from growth in real food products with the predictions from the original model, and then be modified and refined as and if necessary. This strategy relies on the use of simple liquid media cultures and indirect enumeration methods to gather a quantity of data sufficient to allow the construction of a model with consistent parameters, and the subsequent validation of that model in foods, but at fewer values of the independent variables.

2.2.6. Quantity of Data

An absolute lower limit to the amount of data required to fit a particular model is the number of parameters of that model, but for reliability of the estimates there must be many more data points than parameters to be estimated. As a rule of thumb we would suggest that, to fit most of the kinetic models in contemporary use reliably, ten to fifteen data points are required **per independent variable.**

Another problem which may arise as a result of too few data points is in the interpretation of graphs. Fig. 2.2a shows an experimental data set for the growth of *Staphylococcus aureus* as an Arrhenius plot, that is, ln(*rate*) *versus* 1/(temperature[K]). Fig. 2.2b is the same data set but with every second point deleted and the symbols representing the data points made larger. Fig. 2.2c is the same data interpreted as two intersecting straight lines. The fits of the two straight lines are quite good as judged by the correlation coefficient, R^2, but as shown in Fig. 2.2a the fit to the complete data set is better described by a model such as that proposed by Davey (1989b) (see 3.3.4.2.).

The interpretation of curved Arrhenius plots as two intersecting straight lines, and the consequent inference that the process studied changes from one Arrhenius-obeying reaction to another at the temperature indicated by the intersection of those two lines, has often been proposed in the biochemical and physiological literature, for example, Charnock (1978), Kuiper and Kuiper (1978), Schechter *et al.* (1978), Neidhardt *et al.* (1990). It is interesting to note that authors often related the intersection of the two linear portions of the plot to the phenomenon under investigation. Earlier, however, Johnson *et al.* (1954) showed that if the ln (*rate*) axis of an Arrhenius plot is extended, then the data appear to lie on a continuous curve, while the same data plotted with a foreshortened ln (*rate*) axis appear to lie on two intersecting lines. While the present authors would not presume to comment upon the hypotheses proposed, the examples in Figs. 2.2 (overleaf) are used to illustrate that the interpretation of graphical data is a highly subjective exercise and that one must rely on more rigorous statistical methods when analysing data. Particular emphasis should be placed upon the study of the sign of the residuals so that systematic deviations from fitted models can be discerned.

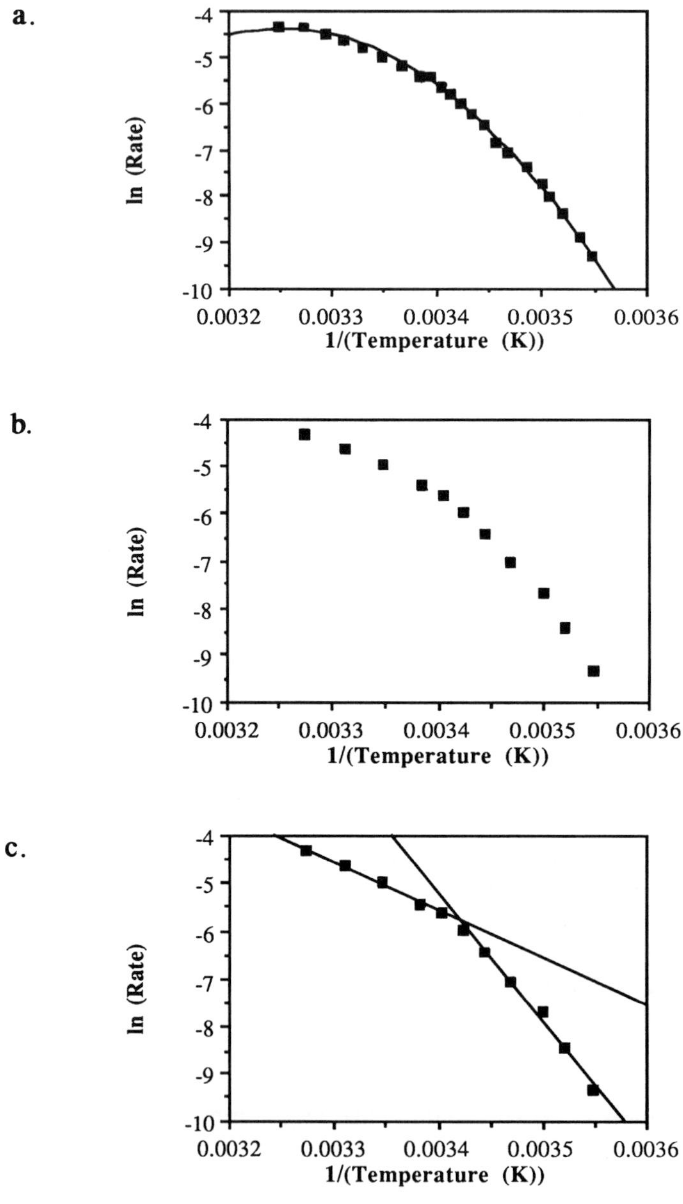
a.
b.
c.
ln (Rate)
1/(Temperature (K))
-4
-5
-6
-7
-8
-9
-10
0.0032
0.0033
0.0034
0.0035
0.0036

d.

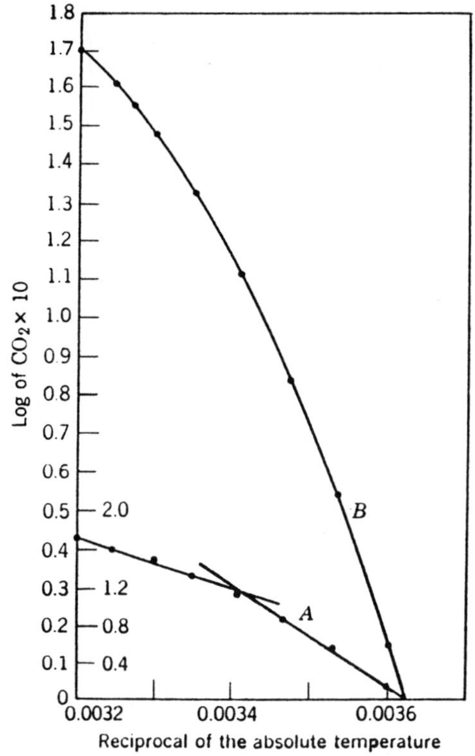

Fig. 2.2 (opposite and above). The figures illustrate the subjective nature of the interpretation of data by graphical means. Fig. 2.2a (opposite) is data for the growth of *Staphylococcus aureus* in laboratory media (data of Ross, unpublished). Fig. 2.2b is the same data as in Fig. 2.2a but with every second datum point omitted. With less points the data might be interpreted as two intersecting straight line responses. The inclusion of such lines on the graph (Fig. 2.2c) further serves to convince the observer. It is, however, misleading. An excellent fit to the data can be achieved by using a second order polynomial (R^2= 0.998) to model the whole data set, rather than using separate fitted straight lines (R^2= 0.995, 0.989). Fig. 2.2d is reproduced from Johnson *et al.* (1954) and shows the rate of carbon dioxide production in alcoholic fermentation as a function of temperature plotted on two different scales. Johnson *et al.* (1954) concluded that when the results are plotted on the expanded scale the function was revealed to be continuous. [Reproduced with permission of John Wiley and Sons].

2.2.7. Summary

The above discussion is a brief introduction to some of the factors that need to be considered when designing the experiments from which to develop a predictive model. It is not an exhaustive discussion; rather it is intended to alert the experimenter to the pros and cons and appropriateness of various methods for specific modelling applications.

2.3. DATA COLLECTION AND ANALYSIS

2.3.1. Definitions

In an environment in which nutrients are non-limiting, a bacterium or population will reproduce and increase in number. If one plots the logarithm of the density of bacteria against time, a characteristic curve results as shown in Fig. 2.3. It can be seen that there is an initial slow rate of increase in log (cell density) rising to an almost constant rate of increase before again slowing to zero rate and finally a negative rate. These four phases of the curve are referred to respectively as the lag; exponential (or often, but inaccurately, logarithmic or log); stationary; and death or decline phases. It is generally accepted that the lag period is a time in which the cells are adjusting their physiology and biochemistry to exploit the environment in which they find themselves, and that during the exponential phase the cells are growing as rapidly as possible in that environment. During the exponential phase cells are said to exhibit **balanced** growth, during which the rate of synthesis of each cell component, whether enzymes, structural molecules, DNA, etc. is so matched as to result in there being no synthesis in excess of the requirement for the production of new cells, that is, the cells' entire metabolic effort is directed to reproduction. In exponential phase all cell components are present in constant proportions and the cells are considered, for practical purposes, to be physiologically identical. As the population continues to increase the accumulation of waste metabolites within the environment becomes sufficiently inhibitory to cause a measurable reduction in the rate of increase in cell numbers. During this phase, conditions may also become sufficiently inhibitory to lead to cell death and lysis, so that the net rate of population increase is further reduced until it is zero i.e. the stationary phase. The stationary phase may be of variable duration but eventually, as further toxins accumulate, the rate of cell death and lysis becomes greater than the capacity of the environment to support cell division, leading to a net decrease in population size recognised as the death phase of the culture. Boyd (1988)

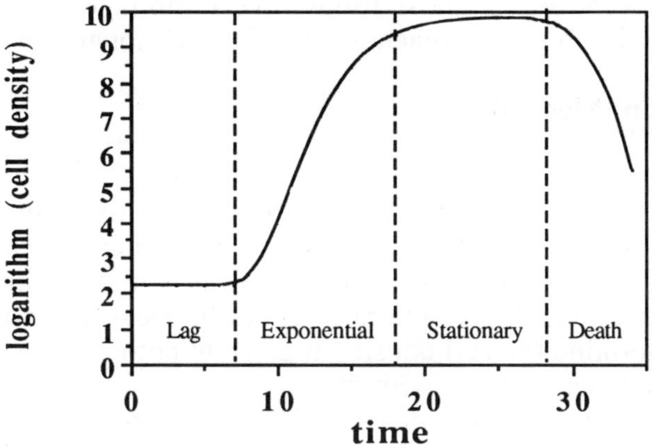

Fig. 2.3 A typical bacterial growth curve showing the increase in population density with time in a closed system. The data are plotted as logarithm (cell density) because the increase is exponential. For bacteria growing in batch cultures the four generally recognised phases of growth are indicated, although Boyd (1988) suggests that pre-exponential and deceleration phases be distinguished also (see text).

identifies two additional periods of growth: an acceleration phase between lag and exponential phases, and a deceleration phase between exponential and stationary phases. The growth and physiology of bacterial populations in batch cultures is covered in most undergraduate microbiology texts, and in greater depth in monographs such as Neidhardt *et al.* (1990).

From the perspective of a bacterium, a unit of food represents a new environment which can be exploited; the bacterium will grow and reproduce in that environment as if in a batch culture. For this reason food microbiologists are interested in the growth kinetics of bacteria in foods, particularly the lag and exponential phases. If the microorganisms in the food can be induced to remain in the lag phase the shelf life of the food can be lengthened. If they achieve exponential growth then the rate of growth under the storage conditions will determine the shelf life of the product. It is useful then for the purposes of predictive microbiology, to be able to model the population growth curve, in particular the duration of the lag phase and the slope of the exponential phase which represents the growth rate of the organism. The stationary phase is of less direct interest because, for most foods, overt spoilage will have occurred before the microbial load reaches that of the stationary phase.

2.3.2. Enumeration Methods

2.3.2.1. Introduction

Probability models require simply the demonstration of growth or production of toxin by the modelled organism to construct a database, but for the construction of kinetic models the *rate* of increase in cell density must be determined. As growth rate cannot be measured directly it is necessary to determine the cell density at discrete points in time and determine the change in cell density with time.

Numerous methods have been described for the enumeration of microorganisms in a given sample. Methods may be classified as **direct** in which individual organisms are distinguished and counted (e.g. direct microscopy, Coulter counter, plate counts), or **indirect** in which some property of the total population (e.g. DNA, dry weight, impedance, optical density) is assayed. A calibration is prepared to relate cell density to the level of the measured property. An example of the need for this calibration is given in 2.5.3. Commonly used methods are briefly described in most undergraduate microbiology texts. Non-culture based methods are discussed in detail by Mallette (1969).

Most workers in predictive microbiology have relied upon either viable (plate) count methods or optical density determinations to derive the growth rate data to build kinetic models. The speed, sensitivity, and potential for automation of impedimetric methods is likely to encourage their greater use in the future.

The MAFF program provides a set of protocols (Anon., 1990) for the production, calculation and recording of data for inclusion in the MAFF Predictive Microbiology database, but also has general relevance. The document aims to ensure consistency of methodology so that data in that database are truly comparable and lead to reliable predictive models.

2.3.2.2. Viable count methods

The advantages of plate counting methods are that they can be made very sensitive, they measure only viable cells, and can be used to determine microbial loads in both real foods and laboratory liquid media. One disadvantage is the possibility that colonies arise from clumps of bacteria rather than individual cells, causing falsely low estimates of cell density in the samples. Consequently, enumeration by viable count methods is reported as the number of 'colony forming units' (CFUs) in the sample. Their chief disadvantage, however, is the amount of labour and materials required to obtain a single determination. It is first necessary to prepare sterile media, diluents, and pipetting equipment. The sample must be serially diluted in order to recover an appropriate number of CFUs on the culture medium, which typically requires the plating out of several dilutions to ensure a 'countable' plate. Plates must be incubated for one or more days to enable CFUs to become (macroscopically) visible, and the counting of colonies is very tedious unless automated systems are available (see 2.3.2.5.). Gibson *et al.* (1987) state that up to 10-15 data points are required to obtain growth curves which can be modelled reliably by sigmoid growth models. This number of samples must be multiplied by the number of conditions to be tested.

Viable count methods are often considered (see e.g. Jarvis, 1989) to have an accuracy of approximately ±0.5 log cycles, although Firstenberg-Eden and Eden (1984) state that the accuracy is perhaps only ±1 log cycle. For these reasons plate count methods usually impose severe restrictions on the quantity of data that can be gathered practically. Consequently, a number of workers have chosen to construct their initial models on data obtained from optical density determinations in laboratory media.

2.3.2.3. Optical density methods

Bacteria suspended in a liquid medium will scatter incident light beams from their original path, in a manner dependent on the concentration, size and shape of the bacteria, the relative refractive indices of the bacteria and the medium, and the wavelength of the incident light. This property is recognised as turbidity and can be used to follow the increase in population density in a liquid medium using a spectrophotometer or a nephelometer as a detector. The relationship between the **turbidity** and the concentration of bacteria is given, over a limited range as:

$$\log\left(\frac{I_{incident}}{I_{scattered}}\right) = -x\,c\,l$$

where:

$I_{incident}$ = intensity of light entering the culture

$I_{scattered}$ = intensity of light scattered by the culture

x = a scaling constant dependent on the characteristics of the suspending medium and the characteristics of the cells

l = length of light path (set by vessel dimensions)

c = concentration of cells

$\log\left(\frac{I_{incident}}{I_{scattered}}\right)$ = **turbidity** of the culture

Thus, for a given vessel, a graph of concentration of cells *versus* turbidity is a straight line of negative slope (-xl). The scattered light can be measured and compared to the incident light with a *nephelometer.*

A similar relationship exists between the proportion of light which passes through the culture, called the **absorbance** or **optical density,** and the cell density:

$$\log\left(\frac{I_{incident}}{I_{transmitted}}\right) = -x'\,c\,l$$

where:

$I_{incident}$, l, and c have the same meaning as above, and x' is analogous to x

$I_{transmitted}$ = the intensity of the light exiting the culture i.e. that which is not scattered so widely that it is not detected by the photometer.

$\log\left(\frac{I_{incident}}{I_{transmitted}}\right)$ = **absorbance**

Other terms used are **transmittance** (T) :

$$T = \left(\frac{I_{transmitted}}{I_{incident}}\right)$$

and **percent transmittance** (% T):

$$\%\,T = \left(\frac{100 * I_{transmitted}}{I_{incident}}\right)$$

The relationship between absorbance and percent transmittance is:

$$\text{absorbance} = 2 - \log_{10}(\ \%\ T\ ) \qquad (2.1)$$

(see Appendix 2A.2)

The advantages of optical density measurements are their speed, simplicity and non-invasiveness which allows many measurements to be made on a single culture. There are, however, limitations. The system usually provides self-consistent data for a particular set of conditions, but is limited to measuring growth in clear liquid media. The relationship between concentration and absorbance/turbidity is only linear over a limited range, corresponding approximately to a tenfold increase in cell numbers. The lower limit of detection is of the order of 10^6 cells/ml. Given that the upper limit of sensitivity is of the order $10^{7.5}$ cell/ml, and that under conditions permitting consistent growth the stationary phase is typically of the order of 10^9 cells/ml, the onset of the stationary phase is not easily measurable, and lag time determinations can only be made on dense populations. These difficulties are illustrated in Fig. 2.4.

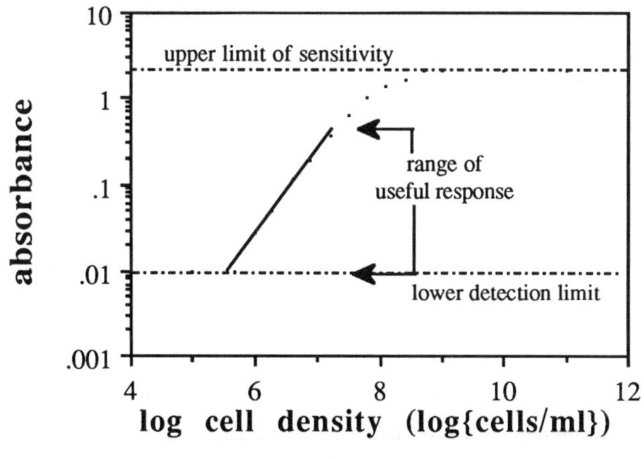

Fig. 2.4. A plot showing the relationship between absorbance and cell density. Though a linear relationship is predicted it is normal to plot the variables as log absorbance *versus* log (cell density), so that the data points are more evenly spaced. Note that the relationship is only linear over a narrow range ($\sim 10^6$ to $\sim 10^{7.5}$), the lower limit of which is determined by the sensitivity of the method.

Furthermore, because optical density values are very dependent on the characteristics of the organism, the medium, and the geometry of the measuring system, the system must be calibrated to absolute cell density if data are to be compared between systems. In predictive microbiology this may not be a problem as the data are usually expressed in terms of kinetic parameters derived from the relative turbidity change with time. It is necessary, however, to verify the range of absorbance values over which the response is linear with cell number. This may be achieved by preparing a very turbid suspension of the organism of interest and evaluating the optical density of a dilution series of the original suspension.

In addition, as most microbiological standards/criteria/guidelines are expressed as numbers of CFU/unit food, one must ensure that models derived by other methods are consistent with plate count results. An example of the need for this is the observation (Hayward, 1990; Ross, *unpublished*) that growth rates calculated from optical density measurements are not always the same as those derived by viable count methods for the same population (see also 2.5.3.). A possible explanation (P. Dalgaard, 1992, *pers. comm.*) is that, at the population densities required for measurable responses using optical density methods, the population is in the late exponential or early stationary phase of growth. Consequently, the measured growth rate will be less than the optimum. Viable count methods, however, have greater sensitivity and are able to measure the growth rate at (lower) cell densities at which the exponential growth rate is maximal.

2.3.2.4. Impedance and conductance

The catabolism of large, uncharged or weakly charged organic molecules e.g. proteins, lipids, carbohydrates, results in the production of smaller, more highly charged products such as amino acids, lactate and acetate. The increase in the concentration of charged molecules alters the electrical properties of the suspending medium, namely reduction of the resistance to current flow. Conductance, which is the inverse of resistance, increases as does the capacitance of the system. If an alternating current is applied to the medium the impedance (AC equivalent of resistance) decreases, whilst the admittance (AC equivalent of conductance) increases. Conductivity and resistivity are, respectively, conductance and resistance per unit length. These changes can be related quantitatively to the initial number and growth rate of microorganisms in a growth medium, which has led to the development of 'impedimetric' or 'conductimetric' growth analysis methods. The impedance changes due to growth consist of both a conductance and a capacitance component. It is considered more convenient to follow the change in conductance because it increases with increasing cell density. Though the method was first described in the late nineteenth century (see Firstenberg-Eden and Eden, 1984), devices based on this phenomenon were not commercially available until the late 1970s and early 1980s. Commercial devices monitor either impedance or conductance changes. Fig. 2.5c is a typical graph of change in conductance *versus* time and population density which shows that there is an initial period during which no change in conductance is observable. This is not the lag time, but rather the *detection time* during which the conductance change due to bacterial metabolism is insignificant compared to the conductance of the medium itself. Firstenberg-Eden and Eden (1984) developed a model (see Appendix 2A.3) which results in the following expression relating population density to detection time:

$$\log C_{BO} = A - B' \cdot t_D \tag{2.2}$$

where:

	C_{BO}	=	population density at time 0
	A	=	a constant dependent on the organism, system and medium
	B'	=	$\log_{10}2$ / generation time
and	t_D	=	detection time

The development of the model assumes that:

i) the generation time is constant, at least in the range of measurement
ii) the generation of ions during the resolution of the lag phase is negligible compared to that since growth commenced, and
iii) a constant amount of ions is produced by each cell in its lifetime.

This expression suggests two methods for the determination of generation times.

If it is assumed that the lag time is independent of inoculum cell density, then the detection time is directly related to the logarithm of the inoculum cell density. By preparing a doubling dilution series in the test medium, the difference in t_D for successive dilutions may be taken as the generation time. Similarly if one plots t_D against $\log_{10}C_{BO}$ the slope of the line of best fit is B' of Eqn. 2.2, i.e:

a.

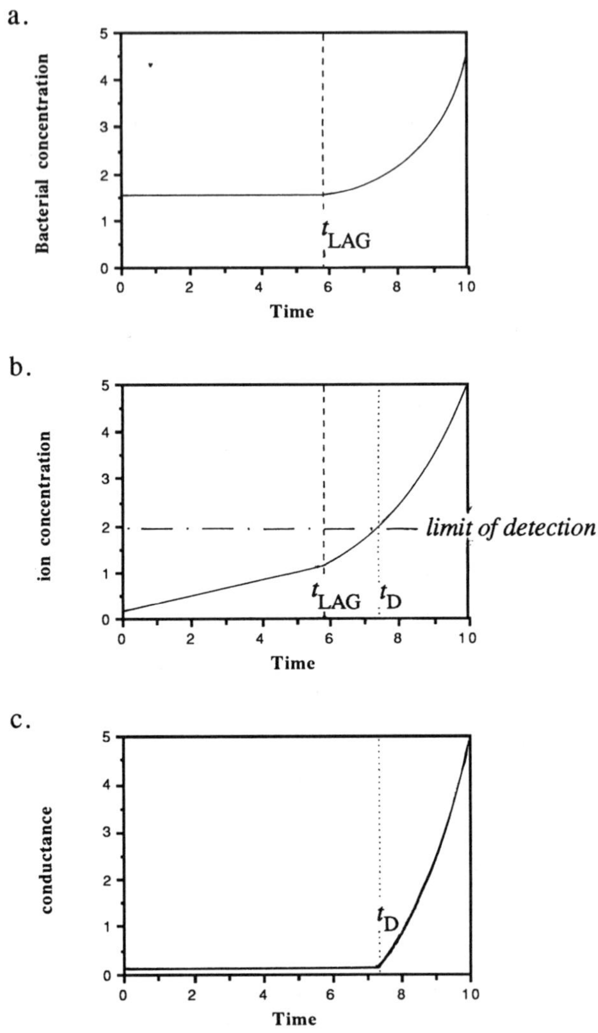

Fig. 2.5 The figures show the relationship between number of bacteria, the evolution of charged molecules due to bacterial metabolism, and the resultant *measured* change in conductance. Note that the detection time is not the same as the lag time, rather, it is the time taken for the change in conductance to reach a measurable level.

[*After* Firstenberg-Eden and Eden (1984)].

$$\text{generation time} = \log_{10}2 / B'$$

There is, to a good approximation, a linear correspondence between the change in conductance and the change in population density during exponential growth (Richards *et al.*, 1978). Therefore generation times may be estimated also by determining the time taken for a doubling of the change in conductance.

Estimation of lag times by impedance methods is more complex, and requires that generation time and cell density at t_D ($N_{[t_D]}$) and t = zero ($N_{[0]}$), or more simply their ratio ($N_{[t_D]} / N_{[0]}$), are known. The lag time may then be determined by extrapolating the population's growth rate back to the initial cell density. Firstly, it is necessary to calculate the number of population doublings ($N_{GEN[t_D]}$) before t_D, which is given by:

$$\text{number of generations} = \log_2(N_{[t_D]} / N_{[0]})$$

$$\therefore \quad N_{GEN[t_D]} = 3.32 \log_{10}(N_{[t_D]} / N_{[0]})$$

$N_{GEN[t_D]}$ is multiplied by the generation time to estimate the time required for that increase in cell numbers. The difference between that time and t_D is an estimate of the lag time, i.e.:

$$\text{lag time} = t_D - \text{g.t.} \times N_{GEN[t_D]}$$

The advantages of the conductance/impedance techniques are their sensitivity and potential for automation. Theoretically the lower sensitivity of detection for enumeration is limited only by statistical considerations and the volume of medium used in the culture test vessel. The method is able to discern changes in conductance due to the metabolism of as few as 10^3 cells/ml (Richards *et al.*, 1978).

2.3.2.5. Automation

A number of devices have been invented which reduce the amount of time and labour involved in plate count methods, e.g. rotating platers, the Colworth "Droplette" (A.J. Seward, Atlas Works, Suffolk, IP33 3PD,

U.K.), image analysis systems, but the method remains extremely time consuming. As we shall demonstrate later, the successful implementation of predictive microbiology will rely on large amounts of kinetic data. Most of these data could be collected by automated systems.

Two automated systems are commercially available for conductance/impedance measurements; the Malthus Microbiological Growth Analyser (Malthus Instruments Limited, Crawley, West Sussex, U.K.) and the Bactometer (Bactomatic, A Division of MTC, Princeton, New Jersey, U.S.A.). Both systems include precisely controlled incubators, the ability to perform hundreds of analyses simultaneously, computer control and memory, and analytical software. Thus the systems can be set up and left without further operator input and the results retrieved at the operator's convenience.

Optical density systems can also be automated. The Biophotorecorder (Advantec. Model TC106. Toyo Roshi International, 6785-A Sierra Crt., Dublin, CA 94568, U.S.A.) is a shaking temperature gradient incubator incorporating a photometer and chart recorder which enables simultaneous continuous monitoring of the growth of six cultures at temperatures from 5°C to 70°C. As the detection device is a photometer, the output of which is an electrical signal, it would be a relatively straightforward task (see 9.2.) to convert this signal into a form which could be recorded and manipulated by a computer with appropriate software to yield a final kinetic result in the desired form.

The Biophotorecorder is based on somewhat dated technology. The 'Bioscreen' [Labsystems (UK) Ltd., Basingstoke, Hants, U.K.] is a more recent invention which *does* monitor the optical density of up to 200 cultures growing in the wells of microtitre plates and also includes software for analysis. The system can automatically dilute and dispense cultures, dispense up to six different components to the wells, and record and analyse the resultant growth curves using operator selected software programs. Whereas the temperature gradient incubator allows growth under identical conditions at many different temperatures to be monitored simultaneously, the Bioscreen allows the simultaneous monitoring of many different conditions isothermally. The Bactometer and Malthus systems currently cost well over US$75 000 and the Bioscreen is quoted as ~US$60 000. The Biophotorecorder is approximately US$15 000.

Criticisms of indirect enumeration methods are that they typically require large numbers of cells to obtain a detectable response, and that the characteristics of the response are often unique to the particular organism and environment under investigation. For example, cell size may change as the salt concentration of the medium changes, and similarly, the rate of change of conductivity may be a function of medium composition. The large number of samples that can be handled by automated systems such as those described above suggest an experimental approach to overcome both of these difficulties. By preparing doubling dilutions of the test organism in each of the test environments, that is, by varying only the initial cell density for each set of conditions, and then incubating a series of four or five such dilutions, the difference in detection time (e.g. specified conductance change, specified optical density) of successive dilutions should represent a generation time. Thus one generates replicates of direct estimates of generation time for each of the environmental combinations studied. The importance of replicated data will be discussed in Chapter 4. In addition, because the only variable in each set of doubling dilutions is the initial cell number, there is no need to calibrate the system for each set of environmental conditions. A further advantage of this approach is that the starting inoculum can be as small as required. Lag time estimates may also be obtained from these data by the method given in 2.3.2.4. if the initial inoculum density, and the cell density at detection time were determined.

2.3.3. Calculating Response Rates and Times

2.3.3.1. Introduction

Traditionally, the lag time and exponential rate of growth have been determined by drawing a line through the exponential phase of the curve or by drawing a tangent to the steepest part of it. The time taken for a doubling of the population, i.e. the generation time, can be determined from the slope of this line. When $\log_{10}$(cell density) is plotted against time, the relationship between slope and generation time is:

$$\text{slope of steepest tangent to exponential growth phase} = \frac{\log_{10}2}{\text{Generation time}} \qquad (2.3)$$

To estimate the lag phase duration, the tangent to the steepest part of the exponential phase is extrapolated back to the density of organisms at time zero, and the time at this point taken as the end of the lag phase. These calculations are shown in Fig. 2.6.

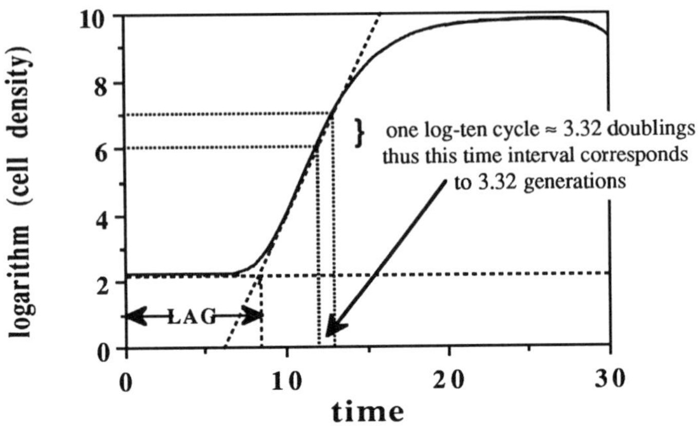

Fig. 2.6 The figure shows a graphical method for the estimation of generation and lag time from a growth curve. The slope of the tangent to the steepest part of the curve estimates the exponential growth rate. The generation time can be calculated from this tangent as the time for a 0.301 unit increase in the log(cell density), i.e. a doubling of the population. The intercept of this tangent with the lower asymptote (i.e. log{cell density} at time = 0) is taken as the lag time.

This method for estimation may be quite subjective, especially where there are a limited number of data points or where the data do not lie on a smooth sigmoid curve. A number of authors have sought to refine the estimation technique. One method (Broughall *et al.*, 1983; Adair *et al.*, 1989) has been to plot the data as log numbers *versus* time and select those points that appear to represent the exponential phase. Linear regression is then applied to these points to optimise the fit of the line through them. The lag time is calculated by substituting the initial log(cell density) into the regression equation which is the mathematical equivalent of the process depicted in Fig. 2.6; the time at this value is taken to be the end of the lag phase. This method is, however, little or no better than the graphical method described above in that a subjective decision still has to be made as to which points represent the exponential growth phase.

2.3.3.2. Sigmoid growth functions

A number of mathematical functions have been proposed which describe sigmoid curves, and which have been used to model the batch growth of microbial cultures. The advantage of such functions is that nonlinear regression techniques can be used to define mathematically the 'line of best fit' through all the data points in an objective and reproducible manner. The fitted equation can be further manipulated to yield the best estimates of the lag time and generation times using all of the data. A disadvantage is that nonlinear regression techniques cannot be applied readily without the aid of a computer and appropriate software.

A brief description of some of these functions is given by Erickson (1970). Causton (1977) describes several sigmoid models and explains their development, and Ratkowsky (1983, 1990) also introduces a number of sigmoid functions and discusses their estimation properties. Models which have been applied to microbial population growth include the Verhulst equation, the logistic function, the Stannard function, and a modification of the Gompertz function.

The logistic, or autocatalytic, function can be developed from first principles for the growth of any population in a non-limiting environment by assuming that the *absolute* growth rate, that is, the change in organism numbers with time, is proportional to the density of organisms at any given time, thus:

$$\frac{\delta \mathrm{N}}{\delta t} = k \mathrm{N} \qquad (2.4a)$$

where:

N = density of organisms (or product)
k = constant of proportionality
t = time

which is simply the equation for exponential growth.

Expressed in another way, the population size increases by a constant *multiple*, given by a constant of proportionality, over a one unit time interval. The *specific* or *relative* growth rate is defined as the increase in population density divided by the cell density at that time, and thus has the value k in the above nomenclature.

In most practical situations, however, environments are not unlimited and the specific growth rate is not constant but declines as a function of the population density (due to depletion of nutrients or build up of toxic metabolites, for example). Thus k in the above equation represents, in some sense, an idealised maximum specific growth rate.

Let the observed specific growth rate be k'. We observe that k' is a function of N, i.e. $k' = f(N)$. Thus we may write:

$$\frac{\delta N}{\delta t} = N f(N) \tag{2.4b}$$

The assumptions upon which the model is based dictate that the maximum specific growth, k, rate must occur when the greatest concentration of nutrients remain, and when the least amount of toxic wastes have accumulated, i.e. when the population is at infinite dilution. Thus at $N = 0$, $k' = k$. Similarly, as the population increases and then reaches the stationary phase, k' decreases to zero. Let us name the maximum value of cell density, a. We can now describe $f(N)$.

The simplest relationship describing the dependence of k' upon cell density is a linear function. We know two points on this line [i.e.$(k,0)$, $(0,a)$], including the y-axis intercept, and using coordinate geometry can now define the dependence of k' upon N:

$$k' = k\left(1 - \frac{N}{a}\right) \tag{2.4c}$$

Combining Eqns. 2.4b, 2.4c we can now derive the logistic function:

$$\frac{\delta N}{\delta t} = k N \left(1 - \frac{N}{a}\right) \tag{2.4d}$$

where:

k, N, and t have the same meaning as previously

and a = a scaling constant that reflects the influence of cell density on the rate of decrease of the rate constant, and which is equal to the maximum population density.

Upon integration and reparameterisation Eqn. 2.4d may be expressed:

$$N_{(t)} = \frac{C}{1 + \exp\{-B(t - M)\}} \tag{2.5}$$

where:

t = time

$N(t)$ = population density at time t

C = upper asymptotic value, i.e.the maximum population density given by $N_{(\infty)}$

M = time at which the *absolute* growth rate is maximal

B/2 = *specific* growth rate at t = M, and where $N(0) \ll C$, B ≈ specific growth rate at $t = 0$

Jason (1983) presented a deterministic model for the monophasic growth of populations in batch culture. The model postulates that the production and growth of each bacterium is accompanied by the generation of a constant mass of end products toxic to the organism, and that the specific growth rate declines in proportion to the ratio of these substances to the dry mass of the nutrient medium when the substrate is non-limiting. Using very detailed and accurate experimental data obtained by impedance measurements he developed the model:

$$n = \frac{2n^*}{1 + \exp\left[(t^* - t) / t_{g'} \right]} \tag{2.6}$$

where

t = time

n = population density at time t

n^* = population density at t^*

t^* = time at which the absolute growth rate is at the maximum

$t_{g'}$ = fastest relative growth rate (i.e. $\frac{\text{change in } n}{n}$), i.e. at $t = 0$ and equals generation time / ln 2

The above expression can be reparameterised (see Appendix 2A.4) to:

$$n = \frac{2n^*}{1 + 2^{\text{number of generations during } (t^* - t)}}$$

Apart from the relationship between cell numbers and the *rate of change* of cell numbers, which Jason (1983) models by the function (using nomenclature consistent with the present discussion):

$$k' = k \frac{C - N_{(t)}}{C - N_{(0)}} \tag{2.7}$$

the development of this model is exactly analogous to that presented by Causton (1977) for the logistic model, but with experimental data to support the assumptions made. Not surprisingly, and as Jason (1983) points out, this model is another parameterisation of the logistic function. It is evident that the parameters of the respective models are analogous, namely t^* and M, $2n^*$ and C, and B and $1/t_{g'}$. Both functions are symmetrical: the logistic function about the point (M, C/2) and the Jason function about (t^*, n*). For these functions the maximum absolute growth rate occurs at the point of symmetry and has the value BC/4 for the parameterisation given by Eqn. 2.5.

The following parameterisation of the Verhulst equation was presented by Broughall and Brown (1984):

$$N_{(t)} = \frac{b}{1 + \frac{b - N_{(0)}}{N_{(0)}} \exp\left\{\frac{-0.693\,(t - L)}{K}\right\}} \tag{2.8}$$

where $N_{(t)}$, and t have the same meaning as in Eqn. 2.5 and

b = maximum population density

$N_{(0)}$ = initial cell density at $t = 0$

$t - L$ = time since resolution of lag phase

K = calculated minimum generation time during exponential growth

For t less than the predicted lag time, N, was set to $N_{(0)}$.

who used it to predict bacterial numbers by substituting into the equation experimentally derived parameter values for the growth of the modelled organism in a variety of environments. This equation can also be reparameterised to give an equation of the same form as the logistic function, Eqn. 2.5 (see Appendix 2A.5).

Note that neither Eqn. 2.5 nor Eqn. 2.6 has an explicit term for the starting inoculum $N_{(0)}$. This is because both models predict that the specific rate of growth is a function of the population size, so that at any inoculum density the specific growth rate will be the same as if the population had grown to that size from any other initial density. It may be argued that some metabolism by the population must occur before the environment is so modified as to inhibit the population growth rate. This issue is reflected in the difference between Eqns. 2.4c and 2.7 used in the development of the logistic and Jason models. Jason's derivation includes the effect of inoculum number, $N_{(0)}$. For a typical maximum population density of 10^9 cells/ml, the values of those expressions differ by less than

1% unless $N_{(0)}$ is of the order of 1/10th of C or greater. In practice, C is usually much greater than $N_{(0)}$ and so the two derivations simplify to the same function. Thus, in practice, there is no need to specify $N_{(0)}$, as both functions model exponential growth only, i.e. after the lag phase has been resolved.

The Gompertz (1825) function was originally formulated for actuarial science for fitting human mortality data but it has also been applied deterministically to organ growth (Causton, 1977). Whereas the logistic type models are based on a linear decline in specific growth rate as a function of population density, the Gompertz function is based on an exponential relationship between these properties. It may be written, in one of its parameterisations, as:

$$N_{(t)} = C \exp\left\{\exp\left[-B(t - M)\right]\right\} \tag{2.9}$$

where

$N_{(t)}$, t , C and M have the same meaning as in Eqn. 2.5. and
B = relative growth rate at M

Each of the above models is formulated as an expression in the absolute density of cells, and gives rise to relationships as shown in Fig. 2.7a. and c. In practice, one is more interested in describing the bacterial growth curve plotted as log (cell density) *versus* time to enable calculation of exponential growth rates and hence generation times. If these expressions are plotted as log (cell numbers) *versus* time they give rise to curves of the type in Figs. 2.7b. and d., from which can be seen more clearly the bases of the models' development. These functions do not model a lag time but only the increase in cell density once exponential growth has commenced. Usually there is a lag period observed before the growth of any inoculum in a new environment. This lag period does not cease abruptly. Rather, there is a gradual transition as more and more of the cells in the inoculum commence growth, i.e. the resolution of lag is not synchronous for all cells in the inoculum. Consequently, when one plots log (cell density) *versus* time to stationary phase one usually observes a sigmoid plot as shown in Fig. 2.3.

Gibson *et al.* (1987) introduced equations similar in appearence to the logistic and Gompertz functions which could be applied to the description of log (cell density) *versus* time bacterial growth curves in terms of exponential growth rates and lag phase duration. Their modification of the logistic model may be written:

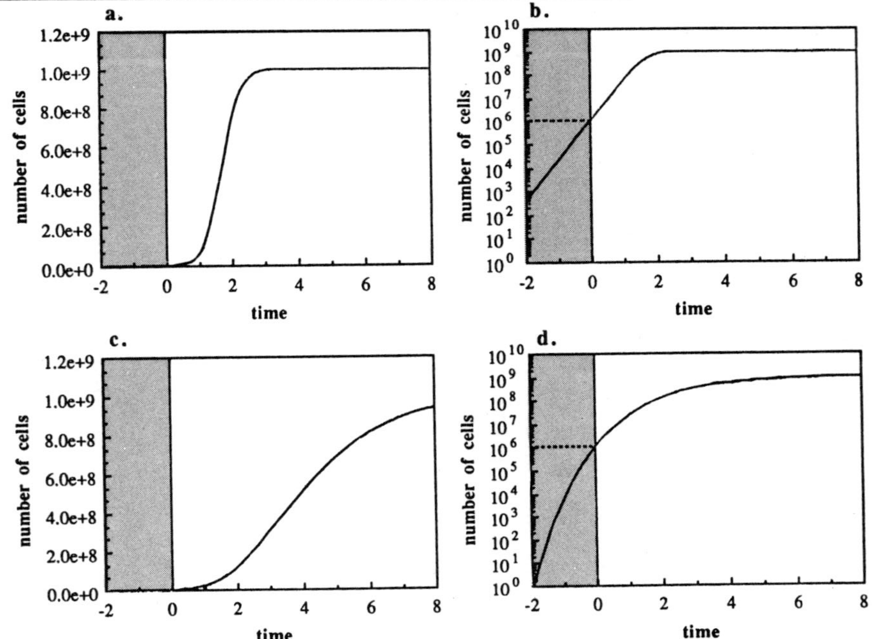

Fig. 2.7 Values for the parameters B and M of a logistic (Eqn. 2.5) and a Gompertz (Eqn. 2.9) function were derived for the growth of a population, having a generation time of 30 min at $t = 0$, from 10^6 to a maximum population level of 10^9 cells/ml. The resultant curves are plotted in Fig. 2.7a and c respectively. Figs. 2.7b and d are the same data presented as semi-log plots, which are more usual for bacterial growth curves. Both Figs. 2.7b and d show clearly that Eqns. 2.5 and 2.9 do not model a lag period, but only the exponential and stationary phases of growth. The figures also illustrate differences in the bases of the models, and that for given values of B, C and M, the fitted curves are independent of $N_{(0)}$, which sets only the starting point on that curve. A smaller inoculum would result in 'more' of the curve, i.e. some of the shaded part on the figures, being observed.

$$\log N_{(t)} = A + \frac{D}{1 + \exp[-B(t - M)]} \qquad (2.10)$$

and their modification of the Gompertz function as:

$$\log N_{(t)} = A + D\exp\{-\exp[-B(t - M)]\} \qquad (2.11)$$

where:
$N_{(t)}$ and t have the same meaning as in Eqn. 2.5
A = value of the lower asymptote [i.e. Log $N_{(-\infty)}$]
D = difference in value of the upper and lower asymptote [i.e. Log $N_{(\infty)}$ - Log $N_{(-\infty)}$]
M = time at which the exponential growth rate is maximal
and where B is related to the slope of the curve at M such that for the Eqn. 2.10:
BD/4 = slope of tangent at M
and where for the Eqn. 2.11:
BD/e = slope of tangent at M

It should be emphasised that the use of log $N_{(t)}$ instead of $N_{(t)}$ as the response variable means that Eqns. 2.10 and 2.11 are not simply reparameterisations of Eqns. 2.5 and 2.9 but are new models. Henceforth we will also refer to Eqn. 2.11 as the 'modified Gompertz' function. Subsequent users of Eqn. 2.11 include the USDA group (Buchanan *et al.*, 1989) and more recently workers at the CSIRO (P.Vanderlinde, *pers. comm.*) and our group at the University of Tasmania.

The additional parameter A is introduced as a location parameter. The mechanistic derivation of the Jason/logistic models only models growth (i.e. once the lag has been resolved) and, consequently, the fitted curve is virtually identical for a given organism/environment system irrespective of $N_{(0)}$. When Eqns. 2.10 and 2.11 are used to model the lag phase as well as exponential and stationary phases it is desirable, for ease of interpretation, to set the lower asymptote at the value of the cell density at $t = 0$. It must be remembered, however, that A really represents the lower asymptote, i.e. log{$N_{(-\infty)}$}, so that if, as may occur in some cases, there is no period of unchanging cell density, the growth curve may in fact be better described by a function for which the value of A is less than $N_{(0)}$. Such a fitted equation nonetheless may predict well the value of the initial cell density $N_{(0)}$. An example of this is shown in Fig. 2.8 (overleaf).

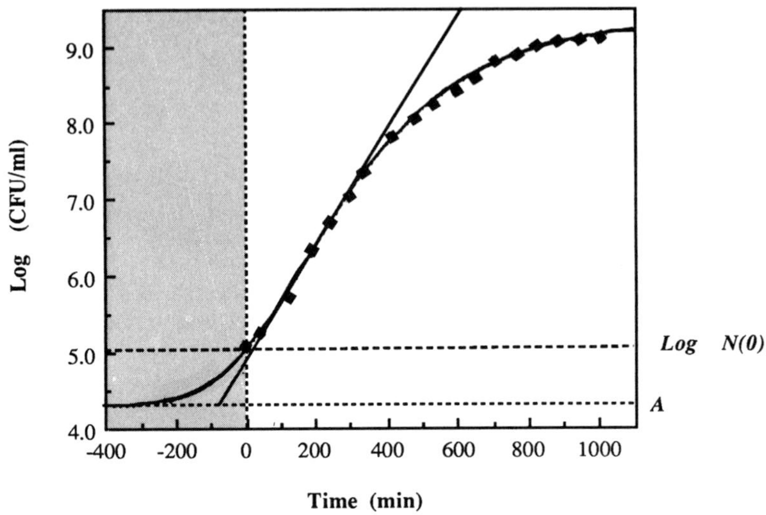

Fig. 2.8 The figure shows a data set (Neumeyer, 1992) for the growth of *Staphylococcus aureus* 3b (♦) and the 'modified Gompertz' function fitted to it (solid line through data points). The 'modified Gompertz' function describes the data well but predicts a value of A which is less than the initial cell density. In fact the fitted equation predicts that log $N_{(0)}$ = 5.04; the observed value is 5.09. This emphasises that A does not necessarily represent log $N_{(0)}$ but is in fact the lower asymptote, log $N_{(-\infty)}$, and also explains why Eqn. 2.16, usually used to calculate lag time, is inappropriate. Using that expression, the lag time is predicted to be -86.75 min, whereas Eqn. 2.15 predicts that the lag time is 14.1 min. The lag time estimated by the traditional 'ruler and eye' method was 14 min.

Part of the reason for this apparent anomaly is that whereas Eqns 2.4 and 2.6 are mechanistic models, Eqns. 2.10 and 2.11 must be considered to be empirical applications of the logistic and Gompertz functions in which $N_{(t)}$ is replaced by log $N_{(t)}$. We are unaware of any mechanistic derivation of Eqns. 2.10 and 2.11.

Stannard *et al.* (1985) also used a sigmoid function to describe bacterial growth curves, but did not indicate whether the response variable was N or log N. The model used was expressed:

$$y = \frac{A}{\left[1 + \exp\left\{\frac{\lambda + \kappa(x + 1)}{\theta}\right\}\right]^{\theta}}$$

where x = time, and A, λ, θ and κ were described as parameters to be fitted.

This model was reparameterised by Zwietering *et al.* (1990), and using nomenclature consistent with preceding equations where possible, may be written:

$$\log N_{(t)} = A + D\left\{1 + \exp\left[-\frac{(b + kt)}{p}\right]\right\}^{(-p)} \qquad (2.12)$$

where:

$N_{(t)}$ and t have the same meaning as in 2.5; A and D have the same meaning as in Eqn. 2.10

and b, k, and p are parameters of the curve which must be estimated

Note that if p = 1, Eqn. 2.12 simplifies to the same form as Eqn. 2.10, that is, Eqn. 2.12 is a more general form of Eqn. 2.10. If Eqn. 2.12 were written with N as the response variable it would be a more general form of Eqn. 2.5, although Jason (1983) shows such a form to be unnecessary.

2.3.3.3. Selection of a sigmoid function to model the growth curve

Whereas the logistic model, Eqn. 2.5, was shown by Jason (1983) to describe accurately the increase in numbers of bacterial populations as a function of time, Eqns. 2.10, 2.11, and 2.12 are empirical and were introduced without rigorous validation of their appropriateness for describing the bacterial growth curve (Fig. 2.3). This issue was addressed by Zwietering *et al.* (1990) who compared the ability of several sigmoid functions, including Eqns. 2.10, 2.11 and 2.12 to model an extensive data set for the batch culture growth of *Lactobacillus plantarum*, and smaller data sets for other bacteria. The comparison was based upon the residual sum of squares between fitted and actual values for each set of growth curve data, but also considered the effect of the number of parameters in each model. The data were fitted in the form of log (cell numbers) relative to that at time = 0, i.e. with the response variable as $\log\left(\frac{N(t)}{N(0)}\right)$. When $\log N_{(0)} \approx A$ (see discussion in 2.3.3.2.) this has the effect of eliminating the parameter A from the equations and setting the lower asymptote to zero, *viz.*:

$$\log\left(\frac{N(t)}{N(0)}\right)$$

$$= \log N_{(t)} - \log N_{(0)}$$

$$\approx \log N_{(t)} - A$$

Using as a yardstick the more general model of Schnute (1981), which, because of its additional parameters encompasses all above mentioned models, Zwietering *et al.* (1990) concluded that in almost all cases the 'modified Gompertz' model could be regarded as the best model to describe the growth data in terms of the goodness-of-fit to the data and its ease of use. The extra parameters of the Schnute model did not provide any useful improvement in goodness-of-fit to the data when compared to the measuring error in the system. Gibson *et al.* (1987) also commented on the usefulness of Eqns. 2.10 and 2.11 for describing microbial growth curves. They reported that many of their growth curves were distinctly asymmetrical and that the logistic function was thus inappropriate.

A new generation of bacterial growth curve models is being developed which have mechanistic bases (Jones and Walker, *in press*; J. Baranyi, *pers. comm.*; R. Whiting, *pers. comm.*) and are reported to allow modelling of all four phases of the bacterial growth curve. At the time of writing, full details of these models were not available and we have no experience of their use. Accordingly, for the remainder of this chapter we will consider only Eqn. 2.11 which, of the models in contemporary use, is the most widely endorsed. Some of the discussion will be relevant to other models also.

2.3.3.4. Derivation of kinetic parameters from sigmoid functions

Once the growth curve has been described mathematically, it can be manipulated to yield the kinetic parameters of the culture such as doubling time, lag time or time to reach a specified level. By differentiating the equation we may derive expressions for the maximum exponential growth rate, minimum generation time and lag time (see Appendices 2A.6. - 2A.8. respectively):

$$\text{maximum exponential growth rate} = \frac{BD}{e} \qquad (2.13)$$

For log (CFU) data fitted by the 'modified Gompertz' function:

$$\text{generation time} = \frac{.8183}{BD} \qquad (2.14)$$

The value of the numerator would, of course, be different if natural logarithms or some other base were used.

The lag time is given by:

$$\text{lag time} = M - \frac{1}{B} + \frac{\text{Log } N_{(0)} - A}{\frac{BD}{e}} \qquad (2.15)$$

This expression differs from Eqn. 2.16 presented by Gibson *et al.* (1987) and subsequently presented by other workers (Buchanan *et al.*, 1989; Buchanan and Klawittter, 1991):

$$\text{lag time} = M - \frac{1}{B} \qquad (2.16)$$

As is explained in Appendix 2A.8, Eqn. 2.15 is a more general and correct expression for the lag time. When $A = \log N_{(0)}$, Eqn. 2.15 reduces to Eqn. 2.16. The use of Eqn. 2.15 would eliminate the reported problem of the prediction of negative values of lag time using Eqn. 2.16.

It should be noted that the calculations leading to the above generation and lag time expressions are not unique to Eqn. 2.11 and could be applied to any other sigmoid function.

It is also possible to adapt optical density data to determination of rate by use of the Gompertz-like function (see Appendix 2A.9).

Buchanan and Cygnarowicz (1990) commented on the somewhat arbitrary nature of methods for estimating lag phase duration and sought a more rigorous approach. They defined the end of the lag phase as the inflection point on the rising part of the curve which can be interpreted as the point of maximum change in exponential growth rate, and which, on a plot of the second derivative of the 'modified Gompertz' function with respect to time, is the maximum. These interpretations are depicted in Fig. 2.9. The time corresponding to this point can be read from the graph or can be solved mathematically by equating the third derivative of the 'modified Gompertz' function to zero. The general solution to this equation is:

$$\exp\left\{ -B\,(t_{LAG} - M) \right\} - \exp\left\{ \frac{-B\,(t_{LAG} - M)}{2} \right\} - 1 = 0$$

where:

B, M	=	the fitted 'modified Gompertz' parameters
t_{LAG}	=	the time at end of the lag period as defined by Buchanan and Cygnarowicz (1990)

From this expression Garthright (1991) showed subsequently that:

$$t_{LAG} = M - 0.9624/B$$

which differs by a small amount only from the more widely used Eqn. 2.16. Again, and as Buchanan and Cygnarowicz (1990) are at pains to point out, this definition of the end of lag phase is not unique for a 'modified Gompertz' function and similar expressions could be derived for any function chosen. While the concept is mathematically elegant, it requires additional computational time and power for a measure that has no greater theoretical validity than the traditional tangent method. The formulae derived by Buchanan and Cygnarowicz (1990) and Garthright (1991) assume that $A = \log N_{(0)}$.

2.3.3.5. Fitting data to the 'modified Gompertz' function

The 'modified Gompertz' function is nonlinear and must be fitted to data by nonlinear regression (see 2.4.2.). It is convenient that the parameters of the equation may be interpreted in terms of features of the data and growth curve, as this enables starting values for the iteration to be readily derived. Provided that the data are obtained during the lag, exponential and stationary phases of growth, the lowest and highest data point respectively can be used as starting values of A and C; B may be estimated by calculating the steepest gradient between any two consecutive points on the rising part of the curve, and M may be taken as the midpoint in time of these two points.

Our experience, and that of others (Buchanan and Cygnarowicz, 1990; Bratchell *et al.*, 1989), suggest that in order to obtain a good fit of the 'modified Gompertz' function to data, and thus to obtain reliable estimates of growth parameters, the quantity and quality of the data are critical.

The fitting process will attempt to derive the sigmoid curve, in accordance with the 'modified Gompertz' function, which most closely fits the data. Thus the data should include values both before and after the inflection point, M. If the data are not well described by a sigmoid curve or are only a segment of one, for example, if there are insufficient data on either side of the inflection point, one should not be surprised if the resulting growth parameter estimates derived from the fitted curve are unrealistic. In the above examples the fitted curve will not be forced towards the upper or lower asymptotes, respectively, as there is insufficient information from which to evaluate them. Bratchell *et al.* (1989) point out that this is less of a problem for the logistic function because it is symmetrical, so that data from one half of the curve can be used to model the other half (see also 2.3.3.2.). Again in this regard it must be remembered that the parameter A does not necessarily approximate log $N_{(0)}$, particularly where the data do not show an initial period of unchanging cell numbers (see e.g. Fig. 2.8). The fitted equation may nonetheless fit the available data well but may be unreliable for predicting the growth curve beyond the range of the data modelled.

Whilst a 'modified Gompertz' function could theoretically be fitted to as few as four points, the consequences of spurious data values are much greater when there are fewer points as Bratchell *et al.* (1989) have reported. They observed that too few points during the period of growth effectively increased the influence that each individual point had on the shape of the fitted curve. Even a single outlying point sometimes considerably distorted the shape of the fitted curve if there were too few other points to offset its influence. They concluded that 10-15 points were necessary to obtain reliable estimates of growth parameters from fitted curves.

It would appear to be sound practice, when using sigmoid functions to estimate growth parameters, to check the goodness of fit of the curve to the data, to set appropriate values for the convergence criterion, and to inspect data visually to ensure that reliable estimates are obtained. Failure to converge is an obvious indication that the data are unreliable or inadequate. Other inadequacies in the data may be less obvious. We undertook a comparison of estimates of the lag and generation times obtained by drawing tangents to hand drawn graphs of several hundred experimentally derived growth curves to those subsequently obtained by using 'modified Gompertz' functions fitted to the same growth data by the

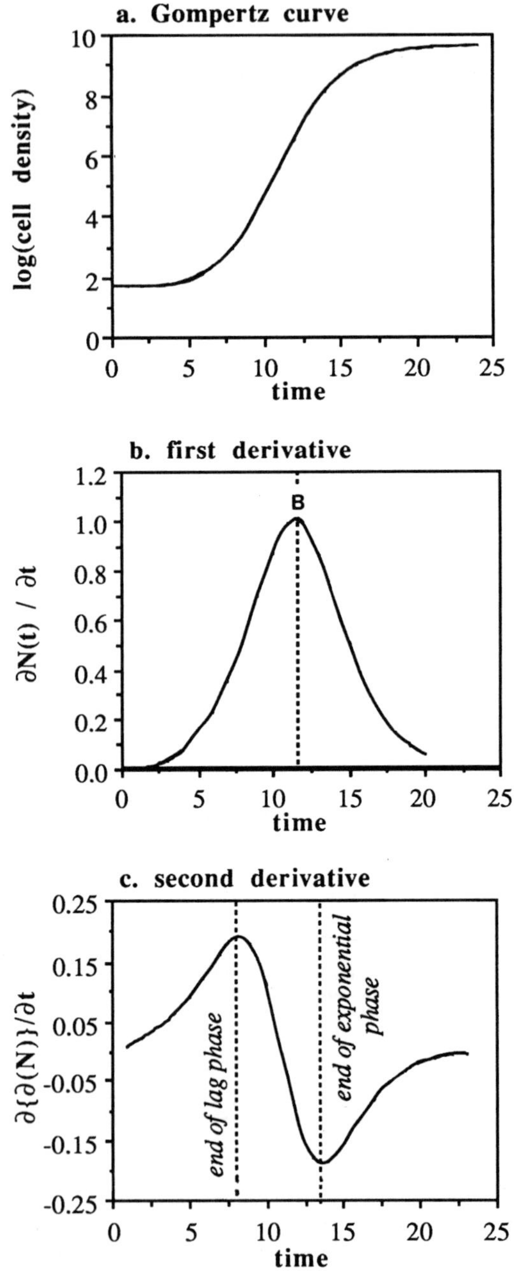
a. Gompertz curve
log(cell density)
time
b. first derivative
∂N(t) / ∂t
B
time
c. second derivative
∂{∂(N)}/∂t
end of lag phase
end of exponential phase
time

Fig. 2.9 (Opposite) Defining the end of the lag phase according to the method of Buchanan and Cygnarowicz (1990). The figure shows the analysis of the slope of Eqn. 2.11, the 'modified Gompertz' function (Fig.2.9a) by taking the first and second derivatives (Figs. 2.9b and c respectively). The maximum of the first derivative, B, occurs at t = M. The maximum and minimum values of the second derivative are the maximum rates of change of the absolute rate, and are interpreted as the end of the lag phase, and end of the exponential phase respectively.

Gauss-Newton method using the nonlinear regression fitting procedure, NLIN, of SAS[1]. The estimates by each method were, in the vast majority of cases, consistent to within ±10% of one another with no consistent trends apparent. As a general rule, we found that data sets which did not converge to the default criterion within less than nine or ten iterations did not give consistent estimates by the two methods, and that typically only five to eight iterations were needed to achieve the convergence criterion for those data sets in which the estimates obtained by the two methods were consistent.

2.4. FUNCTION FITTING

2.4.1. Introduction

The data obtained by experiment must be further processed to fit the data to the model type chosen and evaluate the model's parameters. Generally speaking, the selection of a function to describe a particular response is a statistical exercise, as one is dealing with estimates of the true response and must consider the probability of the response observed being an accurate indication of the true or general response. This point is further discussed later. In contrast, the actual fitting of data to a chosen function is largely devoid of statistical considerations and is, rather, an exercise in algebra.

2.4.1.1. The principle of least squares

The actual process of fitting a function to a data set, that is, to determine the parameter values that best fit the model to those data, is based on the

[1]SAS (Statistical Analysis System) (1987). SAS/STAT Guide for Personal Computers, Version 6 Edition, SAS Institute Inc., SAS Circle, Box 8000, Cary, North Carolina 27512-800, USA.

principle of least squares. This criterion aims to derive parameter values that minimise the sum of the squares of the differences between observed values and those predicted by the fitted model, i.e. the residuals. The application of the least squares principle is not always straightforward, however, and there is now an extensive literature on function estimation (Causton, 1987).

Whilst the principle of least squares is *in itself* largely devoid of statistical concepts, to use it *appropriately* to fit functions to data one must consider some statistical aspects of the response being modelled. In addition, analysis can be applied to the data to determine the statistical properties of the parameter estimates, e.g. confidence intervals. Statistical theory shows that, for independent and identically distributed normal error (see 2.6.), least squares estimation *is* appropriate. When the errors are not normally distributed, then least squares estimation is devoid of statistical 'underpinning'.

2.4.1.2. Stochastic behaviour

Earlier we differentiated between the deterministic and stochastic part of the model. The stochastic part is included to account for the random variation in the response and, if fully specified, would indicate the probability of the measured response falling within a particular range of the response predicted by the deterministic part of the model.

In fitting functions to data by least squares it is generally assumed that the error in the data is homogeneous, that is, the error has about the same magnitude irrespective of the value of the observation. The validity of this assumption can only be tested by having estimates of the error over the range of response variable values. The principle of least squares aims to minimise the sum of the squares of the residuals. Consequently, if the error is not homogeneous, more weight will be given to those points in which the magnitude of the error is large because the greatest reduction in the RSS (sum of squares of the residuals) will be achieved by more closely fitting to those points.

There are a number of methods available to overcome the problem of unhomogeneous errors. One is to use weighted least squares regression analysis. In this method a weighting is applied to individual data points, in inverse proportion to the magnitude of the error variance, for the purposes of estimating the least squares parameter values. Another is to transform the model mathematically into a form in which the response variable does display homogeneous error. Transformation of the model may also be used to improve other properties of the model, such as better parameter estimation in the case of nonlinear models (see 2.6.2.4.).

Causton (1987) argues against the use of weighted regression analysis, however, on the grounds that with modern computing facilities the model can be fitted without undue difficulty in the transformation in which the variance in the response variable is homogeneous, and that "any fudging with weighted regressions is just not worthwhile." The important issue is not the difficulty of fitting the model, however, but that the error behaviour is investigated and correctly specified when fitting the model. Stochastic assumptions, and the consequences of their incorrect specification, are further discussed in 2.6.2.4. and exemplified in Chapter 4.

2.4.2. Regression Analysis

The usual steps involved in fitting *any* function to data by least squares are to:

(i) formulate an expression for the sum of the squares of the residuals
(ii) partially differentiate the expression obtained with respect to each of the parameters, in turn, of the fitted function
(iii) equate the resulting partial derivatives to zero
(iv) solve the resulting set of (simultaneous) equations

These steps are followed no matter what function is being fitted, but the complexity of calculations increases very rapidly as the number of terms in the equation increases. Accordingly, even for the case of a simple linear regression which may be performed on many pocket scientific calculators, the calculations would normally be performed by a computer with appropriate software.

A certain class of functions exists in which the parameters do not appear linearly in the expression of the function. A consequence of this is that when the partial derivatives are calculated and equated to zero, the resulting equations cannot be explicitly solved to generate a single optimal value for each parameter. These are nonlinear models. In some cases, sometimes called intrinsically linear (or nonintrinsically nonlinear) models, the model can be transformed into a linear form which can be used for parameter estimation purposes, although as indicated above this may be an inappropriate strategy if it leads to an unhomogeneous error, or non-normally distributed error, in the transformed response variable.

One way to tackle nonlinear function fitting is by an iterative (stepwise) process in which the parameter values are approximated and substituted into the model. The function is then evaluated and from this a new set of parameter estimates derived so as to reduce the sum of squares of the residuals. The function is then evaluated again, the residuals recalculated using the new parameters, and the process repeated until the difference in the sum of squares between successive estimates (or some function of it such as a ratio) is less than some predetermined value. Non-convergence results if this criterion is not achieved within a user-determined number of iterations.

A number of techniques are available for fitting nonlinear functions to data including the linearisation method (also called the Taylor series or Gauss-Newton method), and its various modifications, and the steepest descent method. These are briefly introduced by Draper and Smith (1981). Fortunately, software is available commercially that incorporates such methods and which can be tailored to suit a particular modelling application. Whilst it may not be absolutely necessary for predictive microbiologists to understand completely the basis of these techniques, it is necessary to be aware of some of the limitations of the methods and the software.

The ability of nonlinear regression to achieve convergence rapidly depends, in part, on good initial parameter estimates. The means by which parameter estimates are obtained will vary from model to model and user to user, but will be aided greatly by having a model or parameterisation for which the parameters are interpretable. For example, in the case of the 'modified Gompertz' function reliable initial estimates can be obtained as explained in 2.3.3.5. If good initial parameter estimates are not used the iteration may converge slowly, or not at all, even though a solution may be available. Another potential problem with iterative procedures is that they may converge to a local, but not overall, minimum sum of squares value. A further problem, associated with the use of commercial software packages, is that the standard errors of the parameter estimates are usually based upon so-called 'large sample-theory', but in most cases in biology and microbiology the sample size is not sufficiently 'large'. This may result in considerable bias in the parameter estimates, underestimation of the true variance in those estimates and, consequently, false confidence in the predictions of the fitted model. Principles of nonlinear regression modelling are discussed at greater length in 2.6.2.

2.5. MODEL VALIDATION

2.5.1. Validation Techniques

No matter how well a model may fit a particular data set, *the true value of a model ultimately relies on how well it can predict microbial responses under novel conditions* that were not specifically tested for in the derivation of that model, i.e. how well it works in the real world. Nonetheless, there are a number of methods that can be used to assess the predictive ability of a model using the data used to generate it, and also to compare the fit of different models to the same data set. A validation of some kind is a useful and necessary part of the modelling procedure.

In general, the basic approach is to divide the complete data set into subsets according to some rational criteria and to use one part of the data set to build the predictive model and to use that model to predict the remaining part of the data set. The predictions can then be compared to the actual data obtained. This may be undertaken by several methods which are reviewed by Draper and Smith (1981).

One approach is to omit sequentially one data point and fit the model with the remaining data. The model is used to predict the value of the omitted point, which is then compared to the actual value observed by calculation of the residual. This process is repeated for each point and a sum of squares of the residuals calculated. Using this method the data can be fitted to different forms of model, and a comparison of the 'goodness-of-fit' to the different models obtained. Other methods discussed by Draper and Smith (1981) are variations and refinements to this approach.

2.5.2. Temporal Validation

In addition to the predictive ability considered above, other properties of the model should be considered. The methods above can be used to validate a model based on data obtained at one time. If the data are collected over a longer time span the parameter estimates may change in time. This can be tested by constructing models from data taken over shorter time spans and considering the values of the fitted parameters as a function of time. If the estimates are not stable in time use of the fitted equation from all the data points would be unwise (Draper and Smith, 1981).

2.5.3. Laboratory to Field

The ultimate test of any model is its ability to predict responses to novel situations. In food microbiology this means the ability of laboratory generated models to predict accurately the behaviour of microorganisms of concern in foods in commercial preparation, storage and distribution. Clearly, validation at this level will have to be performed before predictive microbiology models can be applied practically with confidence. Kilsby (1990, *pers. comm.*) expressed concern at the potential consequences of the premature use commercially of a predictive model. From the perspectives of consumer safety and confidence, and industry acceptance of the predictive microbiology concept, the result of failure of an unvalidated model would be serious, and could perhaps lead to dismissal of the predictive microbiology concept as unreliable.

For models based on data derived from non-food systems, assurance that the model is applicable to food systems is essential, i.e. after the model has been developed it must be tested against observations in real foods. This has been approached in several ways.

Both the MAFF and USDA groups have tested their models' predictions against data from the literature (Buchanan, *pers. comm.*; Roberts, *pers. comm.*; Gibson *et al.*, 1988). Gibson *et al.* (1988) also tested predictions of a model based on the growth of *Salmonella* in laboratory media against growth in slurries of minced defatted pork at seven combinations of temperature and NaCl concentration.

Ross and McMeekin (1991) developed a model based on optical density measurements for the growth of *Staphylococcus aureus* in laboratory media. They took a similar approach to validation as Gibson *et al.* (1988) and inoculated *Staphylococcus aureus* on to fisheries products under 18 conditions of temperature and water activity covering the range of the model. In the course of this validation it was found (Ross, *unpublished*) that although the results of that model were qualitatively correct, the generation times calculated by optical density measurements were of the order of 30% greater than those determined for the same cultures by viable counts (see also 2.3.2.3.). A simple calibration factor was determined and introduced into the optical density based model to enable it to predict growth rates determined by viable counts. Other indirect examples of the validation of laboratory media based models are discussed in Chapters 6 and 7.

2.5.4. Model Maintenance

Draper and Smith (1981) indicate that once a model has been validated and instituted, in practice it is still necessary to set up a procedure for maintaining the model and to recognise when it is becoming obsolete. They recommend that for routine checks standard quality control charts should be adequate but that the model should also be periodically checked by a statistical analyst who may be able to discern more subtle forms of deviation.

2.6. MODEL COMPARISONS

As stated above the most important property of a predictive model is that the model predicts accurately and reliably microbial behaviour in foods. Where competing models perform equally well, that is, within the guidelines for acceptance that have been set, other properties of the model should be considered.

2.6.1. Comparisons of 'Goodness-of-Fit' Between Models

The demonstration of equal performance of models is in itself fraught with difficulty: "The choice between competitors is not straightforward and requires careful comparison of the fitted models or predicted values (Bratchell *et al.,* 1990)". Several methods of comparing predictive kinetic models have been published.

The methods presented by Bratchell *et al.* (1990) are basically pictorial methods that indicate and partly quantify differences between the values predicted from one model and those predicted by another, that is, they serve to highlight *regions of difference*. Some of the techniques presented require sophisticated software and powerful hardware. To use these techniques to compare the goodness-of-fit of different (competing) models one would have to compare the predicted values of each model to the best possible estimate of the true response (e.g. the means of the observed values) in turn, and then devise some procedure to assess objectively their differences. More objective statistical methods are, however, available.

The methods presented in 2.5.1. can also be applied to the objective assessment of the goodness-of-fit of models to data. The best fitting model is that which displays the lowest RSS, although in a statistical sense the model may not be *significantly* better than one with a higher RSS. In addition, models which describe the data well would be expected to give consistent estimates of the model parameters.

Adair *et al.* (1989) directly compared the performance of a 'Schoolfield type' model (see 3.3.3.) with a square root-type model (see 3.2.) for a number of data sets. Both models were expressed with time as the dependent variable. They compared the Mean Square Error of the difference between the observed data and predictions from each of the models (the residuals), that is:

$$\mathrm{MSE} = \frac{\Sigma\{(y_i - \hat{y}_i)^2\}}{n} \qquad (2.17)$$

where:

y_i	=	the observed value (i.e. response time) of the ith data point
$\hat{y}_i$	=	the value of the ith data point predicted by the fitted model
n	=	number of data points

This method was commented upon by Davey (1989a), McMeekin *et al.* (1989), and Zwietering *et al.* (1991) because the comparisons were made in different transformations of the dependent and independent variables, and involved different numbers of parameters. More importantly, MSE is an inappropriate criterion for evaluation of competing models to predict response *times* because the distribution of such time variables is is highly non-normal, typically having a long right-hand tail (see 2.6.2.4., 4.4.).

Zwietering *et al.* (1991) presented a more sophisticated method of evaluating the goodness-of-fit of competing models which considers the sum of the squares of the residuals and the number of parameters of the models being compared. The method compares the fit of the model to the best fit possible, which is that given by a model with the same number of parameters as data points, i.e. each value of the predictor variable is modelled separately. The RSS of this model is calculated from the difference between the *mean* value of the response at each value of the predictor variable and the values of the replicate values of the response at each value of the predictor variable and is a measure of the measuring error in the system. Similarly the RSS may be calculated for a fitted model, but is in this case calculated as the sum of the squares of the difference between the *fitted* value and the replicate observed values of the response. Zwietering *et al.* (1991) described the difference between these values as the lack of fit. RSS_{fit}, the RSS of the fitted model, will always be larger than or equal to RSS_{meas}, the measuring error. If RSS_{fit} is not much greater than RSS_{meas}, then the model is adequate. If RSS_{fit} is much larger than RSS_{meas} the model is inadequate. They evaluated the difference between the lack of fit and the measuring error by the f testing value:

$$f = \frac{(RSS_{fit} - RSS_{meas})/(df_2 - df_1)}{RSS_{meas}/df_1} \qquad (2.18)$$

tested against: $\quad F^{\,df_2 - df_1}_{\,df_1}$

which is more commonly written as $F_{df_1, df_2 - df_1}$, where df_1 is the number of degrees of freedom from the general model and is equal to the total number of data points less the number of different values of the independent variable; df_2 is the number of degrees of freedom of the model being evaluated which is given by the number of datum points less the number of parameters of that model. This procedure is not completely rigorous statistically when applied to nonlinear models, but does at least give an indication of the suitability of different models.

Of those models which Zwietering *et al.* (1991) found to give a statistically acceptable goodness-of-fit, those with fewer parameters, i.e. those more parsimonious, were considered preferable. Parsimony is one of a number of other properties of models that can be used as a basis for comparison. Others include parameter behaviour, parameter interpretability, correct error description, the range of values of the variables, and the ease of use. These issues, though not relevant to all models, are discussed in the following section.

2.6.2. Guidelines for Nonlinear Regression Modelling

The least squares estimators of nonlinear regression models are *biased* and *non-normally* distributed unlike the corresponding estimators in linear regression models, which are unbiased, achieve a minimum variance bound, and are normally distributed (if the errors are normally distributed). Therefore, care must be taken to try to obtain models whose estimators have only a small amount of:

(i) bias
(ii) excess variance above the minimum variance bound, and
(iii) non-normality, as measured mainly by the coefficient of skewness.

Ratkowsky (1983, 1990) has argued that it is a desirable goal of nonlinear modelling to search for models that are "close-to-linear", that is, ones that come close to achieving the properties that are attainable in linear models. 'Far-from-linear' models may have such highly biased estimators for at least one parameter that one can not have much faith that the estimates produced are sufficiently close to the true (but unknown) parameter values. Since the parameters of these models often purport to have physical or biological meaning, grossly biased estimators are clearly undesirable.

Some practical considerations to which modellers should pay attention when engaged in nonlinear regression modelling are:

i) parsimony
ii) parameter estimation properties
iii) range of variables
iv) stochastic assumptions
v) interpretability of parameters

These considerations are discussed in greater detail in the following sections. The reader is also referred to Chapter 3 of Bates and Watts (1988) and Chapters 2 and 10 of Ratkowsky (1990) for further expansion of the ideas presented here.

2.6.2.1. Parsimony

Aim for as parsimonious a model as possible, that is, one with a minimum number of parameters

This accords with the philosophical principle of Ockham's Razor: "*Non sunt multiplicanda entia praeter necessitatem*", which translates "Entities are not to be multiplied beyond necessity". Generally speaking, the greater the number of parameters in a nonlinear regression model, the greater the extent of nonlinear behaviour, that is, the more biased the parameter estimates, the greater their departure from normality, etc. If two models fit data equally well, and one is a two-parameter model and the other a four-parameter model, then clearly the two-parameter model should be the model of choice. There is also a pragmatic justification for parsimony. If we wish to fit any model by the method of least squares, the general procedure (see 2.4.2.) is exactly the same as that for the simple straight line regression except that for nonlinear regression models explicit formulae for the parameter estimates are not obtainable, and iterative methods are required to estimate the parameter values. In addition, the mechanics of calculation increase sharply with the number of parameters. In practice the use of a computer is necessary for almost all problems except when the number of parameters is small. Although Draper and Smith (1981) consider that the necessary calculations could be performed manually for an equation of up to five parameters, an equation of even four parameters would be nearly impossible to solve manually. A computer is necessary for other reasons also, e.g. to calculate fitted values, make normal probability plots with the residuals, calculation of leverage and influence coefficients (Belsley *et al.*, 1980; Cook and Weisberg, 1982) etc.

2.6.2.2. Parameter estimation properties

Given that a good basic model can be found that fits the data well, search for a good reparameterisation of it.

The fact that a model may be found that fits the observed data well does not guarantee that its estimation properties will also be good. For example, consider the Michaelis-Menten model used to describe enzyme reactions widely used by biochemists in the form:

$$v = \frac{v_{max}\,[S]}{[S] + K_m}$$

where

[S]	=	substrate concentration
v	=	reaction velocity
v_{max}	=	a parameter representing the maximum reaction velocity possible
K_m	=	a parameter representing the substrate concentration at which $v = v_{max} / 2$

Despite the good fits that biochemists often achieve using the above model, the least squares estimates of v_{max} are often severely biased, especially when there is a shortage of high substrate concentrations (which would give reaction velocities in the vicinity of v_{max}). Better estimation properties are achieved when the following parameterisation of the above model:

$$v = \frac{[S]}{\theta_1\,[S] + \theta_2}$$

is used (see Ratkowsky, 1986). The 'new' parameters, which are related to the 'old' parameters by the expressions:

$$\theta_1 = 1 / v_{max}$$
$$\theta_2 = K_m / v_{max}$$

give rise to 'close-to-linear' least squares estimators, that is, the estimators are close to being unbiased and normally distributed. It is important to realise that the two parameterisations fit the data in an identical fashion, that is, the 'fitted' values of the response variable are the same.

2.6.2.3. Range of variables

Use as full a range of the explanatory (independent) variables as possible, to obtain as full a range of the response (dependent) variable as possible.

Many problems in nonlinear regression estimation, such as failure to obtain convergence of the iterative process to the least squares estimates, or finding that the estimators are extremely biased, may be attributed to failure to have data over as full a range of the variables as theoretically possible. Thus, as mentioned in 2.6.2.2., part of the reason for considerable bias in the estimator of v_{max} may be attributable to the fact that the largest [S] values may give v values that are too far from v_{max}. Another example involves models for sigmoidal functions such as the Gompertz curve. Data which describe the lower part of the response curve but scarcely go beyond the inflection point must inevitably lead to poor estimates of the parameter representing the asymptote, and may present considerable difficulties in obtaining convergence to the least squares estimates.

Use of a full range of values for the explanatory variable can also help the modeller to distinguish between a good-fitting model and a poor-fitting model. A good-fitting model will tend to give consistent parameter estimates when one or more data points are eliminated in a systematic manner, such as dropping the point with the largest value of the explanatory variable, then dropping the next lowest point and so on. A poor fitting, that is, incorrect model will not give consistent estimates under such an elimination system.

It is generally considered that it is unwise to use a model to extrapolate, that is, to predict responses beyond the range of the data used to generate the model. This applies to all types of models, whether linear or nonlinear. An example of the errors that may result from extrapolation from models, even where the fit of the model to the data is very good, is presented in Appendix 5.2.

2.6.2.4. Correct stochastic description

It is important that the 'error' term (also called 'stochastic' term or 'disturbance' term) be correctly specified, as well as that of the deterministic part of the model.

For example, consider the rational functions model:

$$Y = (\beta + \gamma X) / (1 + \alpha X) + \varepsilon.$$

The deterministic part of the model is the term involving the ratio of two linear functions, viz:

$$(\beta + \gamma X) / (1 + \alpha X)$$

and the stochastic part is expressed by the term ε. It is helpful, though not essential, if ε is 'iidN', that is, independently and identically distributed with a normal (Gaussian) distribution. When this is true, the usual ordinary least squares procedure is justified. If ε is independent and normally distributed, but with a variance that changes as the response variable Y does, weighted least squares regression may then be used. If it is not independent, then more advanced and complicated estimation procedures need to be employed, but these require knowledge of the nature of the dependence.

If the error term is not normal, but belongs to the so-called exponential family, such as Poisson, gamma or binomial, then it is possible to fit the model using iteratively reweighted maximum likelihood estimation (GLIM, 1978). Sometimes, transformation may make the error term more normally distributed. For example, consider the model:

$$Y = \gamma \exp(\beta X) + \varepsilon$$

or its reparameterisations:

$$Y = \exp(\alpha + \beta X) + \varepsilon$$

or:

$$Y = \gamma \lambda^X + \varepsilon$$

Modellers would often consider the transformation obtained by taking logarithms of both sides of the equation, *viz*:

$$\log Y = \alpha + \beta X + \varepsilon'$$

This linearises the deterministic part of the model, $\alpha + \beta X$, but the new stochastic term ε' will be distributed differently from ε. If ε were iidN, the logarithmic transformation would then result in ε' having a smaller variance when Y is large compared to when Y is small. It is important that modellers should take this into account by using a weighted regression analysis with appropriate weights or by fitting a model using GLIM (1978) with 'gamma' error and a logarithmic 'link' function.

If, however, the error term in the original, untransformed, model is 'multiplicative', so that the expression can be written as:

$$Y = \exp(\alpha + \beta X + \varepsilon'),$$

where ε' is the stochastic term, then the logarithmic transformation results in an 'additive' error, allowing the ordinary least squares estimators to be used.

Since it is rare that the form of the stochastic term is known in advance, the assumption is usually tested using the residuals (being the differences between the observed and the fitted values) after fitting the model. Tests based upon the residuals are described in Draper and Smith (1981), Cook and Weisberg (1982) and Belsley *et al.* (1980).

2.6.2.5. Interpretability of parameters

Interpretable parameters, provided that they have good estimation properties, would be desirable.

It is a desideratum that at least some of the parameters in a nonlinear regression model be interpretable. Pragmatically, interpretable parameters often lead to better initial parameter estimates and consequently may simplify the model fitting process (see 2.4.2.). Nevertheless, interpretability of a parameter is no guarantee that its least squares estimator will be close to being unbiased and normally distributed. Consider the Michaelis-Menten model in 2.6.2.2. Biochemists prefer parameters such as v_{max} and K_m to θ_1 and θ_2 because the former are more interpretable, but their estimators are more biased than those of the latter. There are, of course, other parameterisations of the Michaelis-Menten model, such as:

$$y = \frac{X(X_2 - X_1)y_1y_2}{(X - X_1)X_2y_1 + (X_2 - X)X_1y_2}.$$

Here, the two 'new' parameters are y_1 and y_2, corresponding to user chosen values X_1 and X_2, respectively. It is easy to verify that when $X = X_1$, $y = y_1$, and when $X = X_2$, $y = y_2$. The parameters y_1 and y_2 are 'expected-value' parameters, since they are the fitted values corresponding to X_1 and X_2, respectively, and are estimates of the expected values of the response in the underlying population.

Expected value parameters have excellent estimation properties, in general, provided that the reference values X_1, X_2, etc. are chosen to fall within the range of the observed X values of the data set. Clearly, they are also 'interpretable' parameters, as they represent the expected responses corresponding to values of the explanatory variable chosen by the user. Thus, for the Michaelis-Menten model, users can compare values of y_1 and y_2 between data sets and make decisions about whether these parameters differ when the conditions that distinguish the data set also differ. Expected-value parameterisations for a wide range of commonly-used models may be found in Ratkowsky (1990).

Models with interpretable parameters in predictive microbiology include the square root models, (Ratkowsky *et al.* 1982, 1983), which contain the parameters T_{min} and T_{max}, which are notional minimum and maximum temperatures for growth, respectively. These parameters also give rise to close-to-linear least squares estimators, so that the bias in the estimates of these parameters is very small, usually less than 0.1°C, enabling these parameters to be used confidently in classifying microorganisms as psychrophiles, psychrotrophs, mesophiles and thermophiles.

2.6.3. Criteria for the Comparison of Competing Models

From the above discussion it is apparent that the most important quality of a predictive model is that the model describes accurately observed data, and equally importantly that it can model accurately novel data, that is, that it is validated in real products. In addition, the model must encompass as full a range as possible of the variables because it is unsafe to use models to predict responses beyond the range of the data used to generate them. When reliable statistical analysis demonstrates that competing models predict equally well, other characteristics of the model should be compared. These characteristics include parsimony; parameter estimation behaviour; parameter interpretability and correct error description.

A further, almost trivial, consideration in the selection of models is their ease of use. It is only when all other criteria fail to provide discrimination between competing models that this be considered a basis for selection. The ease of use of a particular model will, of course, depend upon one's own environment, resources and background (and possibly prejudices). The factors that may be considered are the ease with which one can 'get a feel' for how the model works so that one is not placing blind faith in the predictions of the model and overlooking potential or real errors, and the amount and complexity of data processing required and software and hardware requirements.

2.7. SUMMARY AND CONCLUSIONS

We have in this chapter reviewed and discussed various strategies used in the development of kinetic models for use in predictive food microbiology. We have also generally discussed the model building process in terms of four recognisable phases. Different researchers favour particular models and we have shown that it is not a straightforward task to compare the performance of models. Furthermore when 'competing' models fit data equally well we must consider other aspects of the model in choosing which to use. Consequently, we have presented criteria for the evaluation of competing models.

APPENDICES 2.

2A.1. Modelling the Growth of Mixed Populations.

This appendix is concerned only with the simple case of a number of strains or species, each of which is growing exponentially.

The expression:

$$\text{relative growth rate} = \frac{d(\log [\text{cell numbers}])}{d\,(\text{time})}$$

is used for any bacterial population.

Consider the situation of two organisms each with a unique specific growth rate in a given environment. Based on the assumption of exponential growth we can derive an expression for the apparent relative growth rate of the mixed population:

$$\text{relative growth rate} = \frac{\log N_{(t + \Delta\text{time})} - \log N_{(t)}}{\Delta\text{time}}$$

substituting into the equation expressions for the increase in numbers of the sub-populations:

$$\text{relative growth rate} = \frac{\log\left(N_0^A \cdot 2^{\Delta\text{time}/GT_A} + N_0^B \cdot 2^{\Delta\text{time}/GT_B}\right) - \log\left(N_0^A + N_0^B\right)}{\Delta\text{ time}}$$

where:

N_0^A, N_0^B *are the initial densities at some time, t, of organism A and B respectively,* GT_A *and* GT_B *are the generation times in the given environment during exponential growth of organism A and B respectively, and* Δtime *is the time interval over which the growth rate is being determined.*

Rearranging:

$$\text{relative growth rate} = \frac{\log\left[\dfrac{N_0^{A}\cdot 2^{\Delta\text{time}/GT_A} + N_0^{B}\cdot 2^{\Delta\text{time}/GT_B}}{N_0^{A} + N_0^{B}}\right]}{\Delta\ \text{time}}$$

$$\text{relative growth rate} = \frac{\log\left[\dfrac{N_0^{A}}{N_0^{A} + N_0^{B}}\cdot 2^{\Delta\text{time}/GT_A} + \dfrac{N_0^{B}}{N_0^{A} + N_0^{B}}\cdot 2^{\Delta\text{time}/GT_B}\right]}{\Delta\ \text{time}}$$

let $\dfrac{N_0^{A}}{N_0^{A} + N_0^{B}} = \rho^{A}$ *and* $\dfrac{N_0^{A}}{N_0^{A} + N_0^{B}} = \rho^{B}$

then

$$\text{relative growth rate} = \frac{\log\left[\rho^{A}\cdot 2^{\Delta\text{time}/GT_A} + \rho^{B}\cdot 2^{\Delta\text{time}/GT_B}\right]}{\Delta\ \text{time}}$$

Using an analogous procedure to the above we may derive a general equation for the observed growth rate of a mixed culture containing n strains or species:

$$\text{relative growth rate} = \frac{\log\left[\rho^{1}\cdot 2^{\Delta\text{time}/GT_1} + \rho^{2}\cdot 2^{\Delta\text{time}/GT_2} + \ldots\ldots + \rho^{n}\cdot 2^{\Delta\text{time}/GT_n}\right]}{\Delta\ \text{time}}$$

where $N_0^{1}, N_0^{2}, \ldots\ldots, N_0^{n}$ *are the initial densities of organism 1, 2,......, n, and* $GT_1, GT_2, \ldots, GT_n$ *are the generation times during exponential growth of organism 1, 2,....,n, respectively and* Δtime *is the time interval over which the growth rate is being determined, and where*

$$\rho^k = \frac{N_0^k}{N_0^1 + N_0^2 + \ldots + N_0^n}$$

where $k = 1,2,\ldots\ldots,n$

Thus the observed rate of growth of the mixed population is a combination of the individual rates and the ratio of the sizes of the sub-populations. The ratio of the sizes of the sub-populations will change with time, however, because those organisms with faster growth rates will increase as a proportion of the total population as a function of time.

Consider a simple situation of two organisms, A and B, initially present in equal proportions in a given environment. Let them have generation times in that environment of 20 and 40 minutes, respectively. Table 2A.1 shows how the apparent generation time of the mixed population calculated over successive 20 minute intervals falls toward that of the fastest growing strain.

Table 2A.1 **The Apparent Generation Time for a Population Comprising Two Sub-populations with Generation Times of 20 and 40 Minutes Respectively**

Δtime (min)	Numbers of A	of B	ρ^A	ρ^B	Total Population Size	Apparent GT (min)
0	1000	1000	0.50	0.50	2000	
10	1414	1189	0.54	0.46	2603	26.3
20	2000	1414	0.59	0.41	3414	25.9
40	4000	2000	0.67	0.33	6000	24.6
60	8000	2828	0.74	0.26	10828	23.5
80	16000	4000	0.80	0.20	20000	22.6
100	32000	5656	0.85	0.15	37656	21.9
120	64000	8000	0.89	0.11	72000	21.4
140	128000	11313	0.92	0.08	139313	21.0
160	256000	16000	0.94	0.06	272000	20.7
200	1024000	32000	0.97	0.03	1056000	20.3

This simple treatment can be extended to predict the behaviour of cocktails of strains/species under a variety of environmental conditions by substitution of appropriate models for the GT terms into the equation. As a general rule the initial rate observed is a weighted mean value of the individual rates under the conditions specified, but rises to that of the fastest growing strain as the duration of incubation increases. In practice, of course, there is a limitation to the amount of growth that can occur.

2A.2. Relationship Between Absorbance and %Transmittance

Transmittance and absorbance are defined:

$$\text{absorbance} = \log_{10}\left(\frac{I_{\text{incident}}}{I_{\text{transmitted}}}\right)$$

$$\text{transmittance} = \left(\frac{I_{\text{transmitted}}}{I_{\text{incident}}}\right)$$

therefore:

$$\begin{aligned} \text{absorbance} &= \log_{10}(\ 1\ /\ \text{transmittance}\ ) \\ &= -\log_{10}(\ \text{transmittance}\ ) \end{aligned}$$

by adding and subtracting $\log_{10} 100$

$$= (\log_{10} 100 - \log_{10} 100)\ -\log_{10}(\ \text{transmittance}\ )$$

and rearranging and evaluating $\log_{10} 100$

$$\begin{aligned} &= 2 - \{\log_{10} 100 + \log_{10}(\ \text{transmittance}\ )\} \\ &= 2 - \log_{10}(\ 100 \cdot \text{transmittance}\ ) \\ &= 2 - \log_{10}(\ \text{percent transmittance}\ ) \end{aligned}$$

i.e.

$$\textbf{absorbance} = \mathbf{2 - \log_{10}(\ \%\ T\ )}$$

2A.3. Development of the Firstenberg-Eden and Eden (1984) Deterministic Model Relating Cell Density, Ionic Concentration and Time

Firstenberg-Eden and Eden presented the following model relating the above entities:

$$C_S(t) = K_B \cdot \frac{t_g}{\ln 2} \cdot \{C_{BO} \cdot 2^{\frac{t - t_L}{t_g}}\}$$

where:

$C_S(t)$	=	concentration of ions at time t
K_B	=	quantity of ions produced by each organism per unit time
C_{BO}	=	density of bacteria at t = 0
t_g	=	generation time of the culture
t_L	=	lag time of the culture
$\{C_{BO} \cdot 2^{\frac{t - t_L}{t_g}}\}$	=	an expression which approximates the number of organisms at time, t, from the initial number C_{BO}, and the number of generations that have occurred since the resolution of the lag phase

The model assumes that the generation of ions during the lag phase is negligible compared to that once growth has reached detectable levels. The basis of the model is the multiplication of the rate of ion generation per organism (K_B) *by the effective cell life-time* $(\frac{t_g}{\ln 2})$, *by the number of life-times* $\{C_{BO} \cdot 2^{\frac{t - t_L}{t_g}}\}$ *which approximates the number of cells at time,* t, *for* $t \gg t_L$. *Note that this model does not consider the reduction in growth rate with increasing population density.*

2A.4. Reparameterisation of Jason's Deterministic Growth Model (Eqn. 2.6)

Jason's Model is expressed (Eqn. 2.6):

$$n = \frac{2n^*}{1 + \exp\left[(t^* - t) / t_{g'} \right]}$$

where n is the number of organisms, t is the time elapsed, $t_{g'}$ is the exponential growth rate, and n^* is the population size at time t^*

Consider the expression:

$$\exp[(t^* - t) / t_{g'}]$$

$t_{g'}$, *the exponential growth rate, is by definition equal to* generation time / ln 2

Thus,

$$\begin{aligned} & \exp[(t^* - t) / t_{g'}] \\ = \; & \exp [(t^* - t) / (\text{generation time} / \ln 2)] \\ = \; & \exp [(\ln 2) * (t^* - t) / (\text{generation time})] \\ = \; & \exp (\ln 2)^{[(t^* - t) / (\text{generation time})]} \\ = \; & 2^{(t^* - t) / (\text{generation time})} \\ = \; & 2^{\text{number of generations during } (t^* - t)} \end{aligned}$$

Thus by substitution Eqn. 2.6 becomes

$$n = \frac{2n^*}{1 + 2^{\text{number of generations during } (t^* - t)}}$$

for $t < t^*$

$$n = \frac{2n^*}{1 + 2^{\text{number of generations during } (t^* - t)}}$$

and for $t = t^*$ $\quad n = n^*$ *as defined*

and for $t > t^*$

$$n = \frac{2n^*}{1 + 2^{-[\text{number of generations during } (t - t^*)]}}$$

$$= \frac{2n^*}{1 + \dfrac{1}{2^{\text{number of generations during } (t - t^*)}}}$$

and approaches the value $2n^*$ *as* $(t - t^*)$ *increases.*

2A.5. Reparameterisation of the Verhulst Equation (Eqn. 2.8)

Eqn. 2.8 is written:

$$N_{(t)} = \frac{b}{1 + \frac{b - N_{(0)}}{N_{(0)}} \exp\left\{ \frac{-0.693\,(t - L)}{K} \right\}}$$

Using nomenclature consistent with the other equations presented it may be written:

$$N_{(t)} = \frac{C}{1 + \frac{C - N(0)}{N(0)} \exp\left\{ \frac{-0.693\,(t - L)}{K} \right\}}$$

Let $\left(\ln\left(\frac{C - N(0)}{N(0)} \right) \right) = C'$

then

$$N_{(t)} = \frac{C}{1 + \exp(C')\cdot\exp\left\{ \frac{-0.693\,(t - L)}{K} \right\}}$$

$$N_{(t)} = \frac{C}{1 + \exp\left\{ \frac{-0.693\,(t - L)}{K} + C' \right\}}$$

$$N_{(t)} = \frac{C}{1 + \exp\left\{ \frac{-0.693\,(t - L) + C'K}{K} \right\}}$$

$$N_{(t)} = \frac{C}{1 + \exp\left\{ \frac{-0.693\left(t - L - C'K/0.693\right)}{K} \right\}}$$

Let L + C'K / 0.693 = M', and ln2/K = B'.
then:

$$N_{(t)} = \frac{C}{1 + \exp\left\{\dfrac{-0.693\,(t - M')}{K}\right\}}$$

$$N_{(t)} = \frac{C}{1 + \exp\,\{-B'\cdot(t - M')\}}$$

which is of the same form as Eqns. 2.5 and 2.6.

2A.6. Derivation of an Expression for Exponential Growth Rate from the 'Modified Gompertz' Function (Eqn. 2.11)

The exponential growth rate may be determined by calculating the slope of the tangent to the modified Gompertz curve (Eqn. 2.11) at the time M, which is the time at which the growth rate is fastest (i.e. the steepest tangent to the curve). Thus we must differentiate the Gompertz expression and calculate the value of the derivative when time = M. Differentiating the Gompertz function as presented in Eqn. 2.11:

$$\log N(t) = A + D \exp\{-\exp(-B(t-M))\}$$

with respect to time:

$$\frac{d(\log N(t))}{d(t)} = D * (\exp\{-\exp(-B(t-M))\}) * \frac{d(-\exp(-B(t-M)))}{d(t)}$$

$$= D * (\exp\{-\exp(-B(t-M))\}) * \{-\exp(-B(t-M))\} * \frac{d(-B(t-M))}{d(t)}$$

$$= BD * \exp\{-\exp(-B(t-M))\} * [\exp\{-B(t-M)\}]$$

By taking the second derivative of Eqn. 2.11 and setting it equal to zero it can be shown that the fastest growth rate occurs at t = M. By substituting t = M into the above we define the maximum exponential growth rate, which can be evaluated as:

$$BD * \exp\{-\exp(-B(0))\} * [\exp\{-B(0)\}] = BD * \exp\{-1\} * 1 = \frac{BD}{e}$$

2A.7. Derivation of Generation Time Expression for Log (CFU) Data Fitted by the 'Modified Gompertz' Function (Eqn. 2.11).

Generation time = *time taken for population to double*

$$= \frac{\log_{10} 2}{\text{exponential growth rate}}$$

substituting the exponential growth rate expression derived in Appendix 2A.6 ;

$$= \frac{\log_{10} 2}{(\text{BD/e})}$$

and simplifiying

$$= \frac{0.3010 * \text{e}}{\text{BD}}$$

$$= \frac{0.8183}{\text{BD}}$$

2A.8. Derivation of a Lag Time Expression for the 'Modified Gompertz' Function (Eqn. 2.11).

The lag time is the time at which the tangent to the curve at M has the value $N_{(0)}$ i.e. the intersection of the line through the initial cell density and the tangent to the curve at M, as depicted in Fig. 2.6. In Appendix 2A.6 we determined the slope of this tangent (i.e. the exponential growth rate). Thus the gradient of the line is $\frac{BD}{e}$.

We can calculate a set of coordinates on this line at time M, from Eqn. 2.11

$$\log N_{(M)} = A + D \exp\{-\exp(-B(M - M))\}$$
$$= A + D \exp\{-\exp(0)\}$$
$$= A + D \exp(-1)$$
$$= A + \frac{D}{e}$$

We may now describe the tangent to the Gompertz function at M:

$$\log N' = \frac{BD}{e}(t - M) + \left(A + \frac{D}{e}\right)$$

To determine the lag time we wish to evaluate this function at the point (**L**, $\log N_{(0)}$) *where* **L** *is the lag time i.e the intersection of the initial value of the cell density with the tangent which is the equivalent of the process described in 2.3.*
Thus we set

$$\log N_{(0)} = \frac{BD}{e}(\mathbf{L} - M) + \left(A + \frac{D}{e}\right)$$

$$\therefore \quad \frac{\log N_{(0)} - \left(A + \frac{D}{e}\right)}{\frac{BD}{e}} = \mathbf{L} - M$$

$$\mathbf{L} = M - \frac{1}{B} + \frac{\log N_{(0)} - A}{\frac{BD}{e}}$$

Note that where $\log N_{(0)} = A$ *this reduces to the equation published by Gibson* et al. *(1987) and others (Buchanan* et al. *1989; Buchanan and Klawitter, 1991).*

From Eqn. 2.11 $\log N_{(0)} = A + D \cdot \exp[-\exp(BM)]$ *which upon substitution into the above expression gives:*

$$\mathbf{L} = M - \frac{1}{B} + \frac{e}{B\left(\exp\left(\exp(BM)\right)\right)}$$

which can be written more simply as:

$$\mathbf{L} = M - \frac{1}{B}[1 - \exp\{1 - \exp(BM)\}]$$

2A.9. Using the Gompertz Function with Optical Density Data.

The generation time of a population can be estimated from the time taken for the absorbance of the culture to double (see 2.3.2.3.) It is most convenient to plot log (absorbance) versus *time, however, because a population doubling corresponds to a log(absorbance) increment of 0.301. For determining generation time it may be more convenient still to measure and plot the increase in population as change in %transmittance which, as we will demonstrate below, increases linearly as population density increases logarithmically.*

If the decrease in percentage transmission is plotted against time it is found to be a sigmoid curve which can be described by a Gompertz function. This is analogous to fitting the Gompertz function to log CFU data, but the parameters now have different meanings as follows:

- *A* = *lower limit of detection of the spectrophotometer or %transmittance of the initial microbial load*
- *B* = *maximum rate of change of %transmittance*
- *M* = *time at which rate of change of %transmittance is maximal*
- *D* = *difference between the lower and upper limits of sensitivity of the spectrophotometer*

The generation time can be estimated from the Gompertz parameters fitted to %transmittance data by introducing a calibration factor into Eqn. 2.13, in the same manner as was shown for log CFU data in Appendix 2A.7.

We know that an increment of log (absorbance) of 0.301 corresponds to a population doubling (see 2.3.2.3.). An algebraic solution of the correspondence between an increment of 0.301 in log (absorbance) and %transmittance shows the solution is dependent upon the value of %transmittance. However, plotting log (absorbance) versus *%transmittance reveals that there is a linear correspondence between these measures over a narrow, but useful range, as shown in Fig. 2A.9.1:*

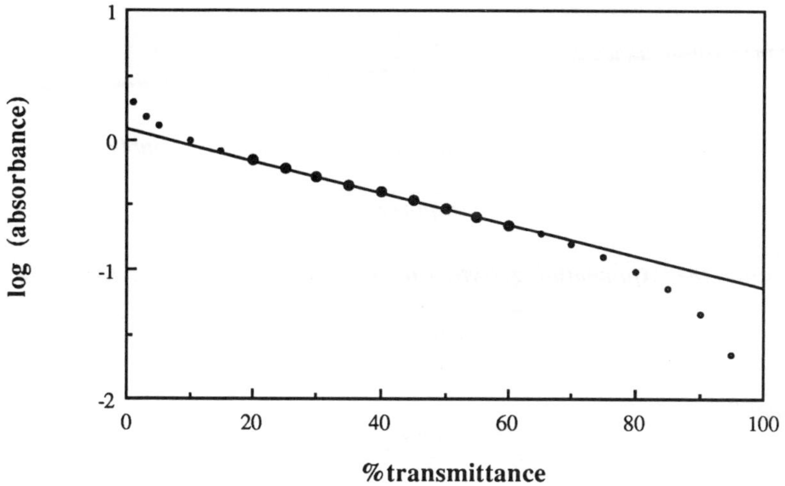

Fig. 2A.9.1 Calibration curve for generation time using %transmittance data. For the range 20 - 60 %transmittance:

log (absorbance) = 0.089525 - 0.012293 x %transmittance ($R^2 > 0.999$)

From the regression line fitted to the data over the range 20 to 60 %transmittance it is possible to derive a calibration value:

$$\textit{slope of regression line} = -0.012293$$

$$= \frac{\textit{change in log (absorbance)}}{\textit{change in \% transmittance}}$$

log (absorbance), thus:

$$0.012293 = \frac{0.301}{\text{change in \% transmittance}}$$

$$\text{change in \%transmittance} = \frac{0.301}{0.012293}$$

$$= 24.5$$

$$\text{Generation time} = \text{time taken for population to double}$$

$$= \frac{24.5}{\text{exponential growth rate}}$$

substituting the exponential growth rate expression derived in Appendix 2A.5:

$$= \frac{24.5}{(BD/ e)}$$

and simplifiying

$$= \frac{24.5 * e}{BD}$$

$$= \frac{66.6}{BD}$$

where B and D are the Gompertz parameters estimated by fitting the function to %transmittance data.

3.

MODELLING TEMPERATURE EFFECTS

The previous chapter was concerned with the question of how one might summarise isothermal bacterial growth into one or more single measures, among which are, for example, specific growth rate, lag phase duration, and generation time. The present chapter is devoted to how those measures change with changing temperature. It will be seen that there are two major classes of models that have been proposed for this purpose, the Arrhenius-type and the Bělehrádek-type. Various equations that have been proposed within each of these classes are presented here.

3.1. INTRODUCTION TO ARRHENIUS- AND BĚLEHRÁDEK-TYPE MODELS

Two types of model have been proposed to describe the effect of temperature only on microbial growth. As indicated in the preface, those based on the Arrhenius equation have gained prominence in microbiological studies to the extent that Bělehrádek-type models were almost totally unrecognised in microbiology for a period of ~40 years to ~1982. This despite the fact that they had been widely used in other biological sciences particularly marine biology. A summary of the earlier work was presented by Ross (1987); more recent usage is discussed in Appendices 4A.5 and 4A.6.

The empirical rediscovery of the Bělehrádek power function in microbiology began in 1973 when Olley and Ratkowsky (1973a, b) utilised a relative rate curve (see 6.2) which, as explained in Chapter 1, described psychrotrophic spoilage of fish and other proteinaceous foods. It is claimed, though the literature is not clear on this point, that bacterial spoilage does not begin until *rigor mortis* is resolved. Time to resolution of *rigor* is dependent on temperature, pH, glycogen reserves and adenylate charge. Olley (1983) explained how a reconciliation of the rates of *rigor mortis* resolution and the psychrotrophic spoilage curve for proteinaceous food was found in the work of Ohta and Hirahara (1977). These authors used the same relative rate concept to describe the rate of degradation of nucleotides, k, in carp, and also expressed their results as a curvilinear rate equation:

$$k = 0.318 + 0.0455t + 0.0053t^2 \qquad (3.1)$$

or by a square root function:

$$\sqrt{k} = 0.0656t + 0.518 = 0.0656[t - (-7.9)] \qquad (3.2)$$

where t is temperature in °C.

Ratkowsky *et al.* (1982) applied this square root transformation to describe the effect of temperature on bacterial growth rate and Bremner *et al.* (1987) indicated its validity for the rate of production of *post mortem* metabolites. The effect of superoptimal temperatures was described by Ratkowsky *et al.* (1983) by adding an exponential term to the simple square root equation.

The square root temperature response has been verified by many authors for bacterial growth and has become generally known as the 'square root' or 'Ratkowsky' equation. However, Ross, quoted by Olley (1986), was the first to point out that this was equivalent to a Bělehrádek temperature function with the exponent having the value 2 (see Eqn. 3.6).

One difference between the two types of model lies in the way in which the response variable is transformed. The Arrhenius equation transforms the dependent variable using the logarithm and transforms the independent variable using the reciprocal. This contrasts with determination of a power function in Bělehrádek-type models in which the exponent is obtained from a log-log plot of rate *versus* temperature. Further, Bělehrádek recognised the need to introduce the concept of 'biological zero' rather than relating biological activity to absolute zero.

In this chapter, we confine our attention to the effect that changing temperature has on the growth of bacterial cells. However, even in an isothermal situation, the growth rate of the cells of a single, pure strain of a bacterium may depend markedly on various factors that can be growth-limiting, such as water activity, pH, availability of nutrients, etc. Hence, in order to put the modelling of temperature effects on a comparable basis, where each culture, irrespective of the bacterial strain involved, has an equivalent set of conditions, we consider only the case of a bacterial culture grown in an unchanging environment in which nutrients are non-limiting. In such circumstances, growth is said to be unrestricted (Ingraham *et al.*, 1983). In Chapter 5 we will consider the effects of such additional factors as water activity and pH and attempt to describe the combined effects of these factors.

3.1.1. Unrestricted Bacterial Growth

The key assumption for unrestricted bacterial growth is the consideration that the rate of increase of the cells of a culture is proportional only to the number of cells present at a particular time, in an isothermal situation. Thus, if N denotes the concentration (number of cells in a given volume) and ϕ the time, the proportionality assumption leads to the differential equation:

$$dN/\,d\phi = k\,N \qquad (3.3)$$

where the constant of proportionality k is used as an index of growth rate for the bacterial strain and is called the specific growth rate. The solution to Eqn. 3.3 for any pair of time ϕ and concentration N is given by:

$$\ln\,(N/N_o) = k(\phi - \phi_o) \qquad (3.4)$$

where ϕ_o is a reference time at which the corresponding cell concentration is N_o. Conveniently, ϕ_o can be chosen to be zero at some measurable initial concentration N_o, but any pair (ϕ_o, N_o) values serves the same purpose. If ϕ_o is set at zero at the time when N_o is measured, Eqn. 3.4 can be expressed as:

$$N = N_o e^{k\phi} \qquad (3.5)$$

where e is the base of natural (Naperian) logarithms. The form of Eqn. 3.5 explains why bacterial growth is often described as following an 'exponential' law.

Although the above discussion was framed around the cell concentration N, the same values of k tend to be obtained if some other extensive property of the growing bacterial system, such as mass, protein or DNA, is measured in place of N, under the specified culture conditions.

If the conditions apply for the exponential law to be valid, a series of specific growth rate constants k can be obtained, one for each temperature t. Since k is constant only at a given temperature t, the question arises as to the manner in which k changes with changing temperature. An examination of various proposed relationships follows under two main headings, *viz.* (i) Bělehrádek-type models, and (ii) Arrhenius-type models.

3.1.2. Bělehrádek-type Models

Although Bělehrádek had earlier published equations for biological rate processes (see Ross, 1987), the form now most widely used was first presented (Bělehrádek, 1930) as:

$$y = \frac{a}{(t - \alpha)^b}$$

where:

y	=	development time
t	=	temperature
α	=	a 'biological' zero
a, b	=	fitted parameters

In modern notation, the Bělehrádek-type models can be written as expressions of the general form:

$$k = a(t - t_o)^d \qquad (3.6)$$

where:

k = rate

and a, d and t_o are parameters to be fitted.

The term t_o is mathematically equivalent to the constant α in the notation of Bělehrádek.

Bělehrádek (1930, 1935) regarded t_o as a 'biological zero', that is, as a temperature at which (and, of course, below which) no growth is possible. Experiments at that time were not sophisticated enough to distinguish between the lowest temperature at which an organism could be perceived to grow within the time scale of the experiment, and the intercept of the mathematical model. The confusion which this has caused will be considered in 3.2.2. Above t_o, the growth rate parameter k, which may or may not be a specific growth-rate constant in an exponential growth situation such as in Eqns. 3.3 - 3.5, increases with increasing difference between t and t_o depending upon the magnitude of d, a positive constant. The higher the value of d, the more rapidly the value of k would increase for a given temperature difference t-t_o. Clearly, the magnitude of the parameter a, also a positive constant, will depend upon which extensive property of the system k is derived from, and the units thereof. Given a reference temperature t_{ref}, and the rate constant k_{ref} corresponding to it, a consequence of the Bělehrádek-type relationship is the expression:

$$k = k_{ref}\left[\frac{t - t_o}{t_{ref} - t_o}\right]^d \qquad (3.7)$$

so that a rate at any temperature in the suboptimal range can be evaluated from known values of t_o, d and k_{ref} (corresponding to the user's choice of t_{ref}). Alternatively, the rate at any temperature in the suboptimal range can be evaluated from knowledge of t_o, d and a.

3.1.3. Arrhenius-type Models

The second major grouping of models relating k to temperature is that of the Arrhenius-type models and their extensions. The thermodynamic relationships underlying these models date back to the work of van't Hoff (1884) and Arrhenius (1889) on chemical reactions. The integrated form of the underlying differential equation results in the following empirical expression:

$$k = A\exp(-E_a/RT) \qquad (3.8)$$

where:

k is the specific reaction rate constant,

E_a is the so-called 'activation energy' of the reaction system,

T is the absolute temperature,

R is the universal gas constant: 0.001987 kcal mol^{-1} K^{-1} = 8.31 J mol^{-1} K^{-1},

and A is the so-called 'collision factor'.

In bacterial growth systems, microbiologists have replaced k by a temperature characteristic μ or simply have interpreted k as a specific growth rate constant, as in Eqns. 3.3 - 3.5.

We make no attempt here to justify one or the other of the class of models, the Bělehrádek-type or the Arrhenius-type, as 'theoretical' *versus* 'empirical' models. Although the Arrhenius model, or at least its successor grounded in the theory of absolute reaction rates, has a firm theoretical basis when applied to chemical reactions involving collisions between simple molecules, no such justification exists for its application to complex reactions involving enzymes, as is the case for bacterial growth. Hence, in this treatment we would view all models to be described in the following sections as empirical models and determine in Chapter 4 the model (or models) which best fit(s) the data and provide(s) a useful tool for summarizing the way in which the specific growth rate constant changes with temperature.

3.2. BĚLEHRÁDEK-TYPE RELATIONSHIPS

3.2.1. The Spencer-Baines Model

One of the earliest attempts to express the relationship between k and temperature was that of Spencer and Baines (1964), who were concerned with the chemical, bacterial and organoleptic spoilage of codfish at temperatures between 0 and 25°C. Their model was expressed as:

$$k = k_o(1 + ct) \tag{3.9}$$

where:

k_o is the spoilage rate at the reference temperature 0°C,
t is the storage temperature in degrees Celsius,
and c is a parameter representing the temperature coefficient.

That is, the specific growth rate constant is a linear function of temperature in that model. Eqn. 3.9 is a special case of the general Bělehrádek-type expression given by Eqn. 3.6 with the exponent d = 1. With d = 1, Eqn. 3.6 is simply:

$$k = a\,(t - t_o)$$

which is a linear function of temperature, as is Eqn. 3.9, although the reference temperatures may differ. The Spencer-Baines model is probably the simplest special case of Bělehrádek's general form.

Spencer and Baines (1964) claimed that the temperature coefficient c was constant between -1 and 25°C, but Olley and Ratkowsky (1973a) showed that it was approximately constant at a value $c = 0.24$ only between 0 and 6°C, and that above 6°C the value of c increased, rendering their approximation to be useful only for a narrow temperature region.

3.2.2. The Square Root Model for Suboptimal Growth Temperatures

Far more successful than the Spencer-Baines model (Eqn. 3.9) has been the simple square root model of Ratkowsky *et al.* (1982):

$$\sqrt{k} = b(T - T_{min}) \tag{3.10}$$

which is intended to be applied only to the low temperature region, that is, for temperatures ranging from the minimum temperature at which observable growth can occur to that just below the optimum temperature for growth. Ratkowsky *et al.* (1982) used the notation T_o but changed this to T_{min} (Ratkowsky *et al.*, 1983) to avoid the term being interpreted as the rate at 0°C. The optimum temperature is the temperature at which the specific growth rate constant is a maximum and T_{min} is estimated by extrapolation of the regression line derived from a plot of $\sqrt{k}$ *versus* temperature to the temperature axis. By definition, the growth rate at this point is zero. T_{min} is usually 2-3°C lower than the temperature at which growth is actually observed and has been described as a **notional, notational**, or **conceptual** temperature. Despite repeatedly emphasising this point (Ratkowsky *et al.*, 1982; McMeekin and Olley, 1986; McMeekin *et al.*, 1987), the difference between T_{min} and the minimum temperature at which growth occurs continues to be ignored (Neidhardt *et al.*, 1990; Zwietering *et al.*, 1991), although it has been well understood by Heitzer *et al.* (1991).

Bělehrádek (1935) applied Eqn. 3.6 to a number of animal species and found a range of d values. Bělehrádek tended to use the lowest temperature at which growth had been observed as his 'guesstimate' of t_o. Ross (unpublished results) has found, using his extensive data set for the growth of *Staphylococcus aureus* 3b, that d is linearly related to T_{min}. Thus any error in the estimation of t_o will cause variation in the value of the exponent d. Ratkowsky *et al.* (1982) found that Eqn. 3.10 applied to 43 strains of bacteria for which there were available growth data. To date, there are no known exceptions to the universal applicability of Eqn. 3.10 in the low temperature region for any strain of bacteria or mould. If one squares all terms in Eqn. 3.10 it is clearly an example of the general Bělehrádek relationship (Eqn. 3.6), with $d = 2$. Since the difference in two Celsius temperatures is identical with the difference of the same two temperatures expressed in degrees absolute (Kelvin), that is:

$$t - t_o = T - T_{min}$$

the general Bělehrádek form (Eqn. 3.6) can just as validly be written as:

$$k = a(T - T_{min})^d \qquad (3.6a)$$

The question of which of the equivalent forms:

$$k = a(T - T_{min})^2 \qquad (3.11)$$

or Eqn. 3.10:

$$\sqrt{k} = b(T - T_{min})$$

where $b^2 = a$

is more suitable for data fitting and interpretation rests on statistical considerations which will be discussed later. It suffices here to say that errors in $\sqrt{k}$ are generally more homogeneous with changing temperature than errors in k (Ratkowsky *et al.*, 1991), so Eqn. 3.10, the 'square root' model, is favoured above Eqn. 3.11 for statistical analysis.

3.2.3. The Four Parameter Square Root Model for the Full Biokinetic Temperature Range

The four parameter square root model:

$$\sqrt{k} = b(T - T_{min})\{1 - \exp[c(T - T_{max})]\} \qquad (3.12)$$

is an extension of Eqn. 3.10 which enables growth data obtained over the whole biokinetic temperature range to be modelled (Ratkowsky *et al.*, 1983). Just as T_{min} is the 'biological zero' at the low end of the temperature range, T_{max} is the 'biological zero' of the high end of the scale. The parameter c, as with b, has no interpretable meaning, being present merely to enable the model to fit the data for temperatures above the optimal temperature. Eqn. 3.12 was successfully applied to data for 29 strains of bacteria, comprising 16 from previous unreported isolates, 12 from the work of Mohr and Krawiec (1980) and one from the work of Reichardt and Morita (1982). Graphs of $\sqrt{k}$ *versus* T are given in Ratkowsky *et al.* (1983) for all 29 strains. The form of Eqn. 3.12 is such that the effect of the term:

$$\{1 - \exp[c(T - T_{max})]\}$$

is negligible when T is much less than T_{opt}, the temperature at which k is a maximum. In the low temperature region, Eqn. 3.12 thus reduces to Eqn. 3.10. At temperatures above T_{opt}, the above term is very influential, dominating the growth rate and causing it to fall sharply to a zero growth rate at $T = T_{max}$.

Once the four parameters b, c, T_{min} and T_{max} of Eqn. 3.12 have been estimated, the optimum temperature for growth, T_{opt}, can be found by solving the equation:

$$c(T_{opt} - T_{min}) = \exp[c(T_{max} - T_{opt})] - 1$$

The set of temperatures T_{min}, T_{max} and T_{opt} are often referred to as the 'cardinal temperatures' characterizing growth. Predicted cardinal temperatures, including those obtained from fitting the 29 strains reported in Ratkowsky *et al.* (1983), are given in Table 3.1. A question arises as to what, if any, is the natural order in which to list the strains in that table. The choice used in the construction of Table 3.1 was based upon increasing values of T_{min} to order the listing, as Ratkowsky *et al.* (1982) suggested that the T_{min} value was an intrinsic property of an organism and useful to categorise it as a psychrophile, psychrotroph, mesophile or thermophile. As with other arbitrary categories in biology, there are no clear cut distinctions between the temperature groups but rather a continuum of T_{min} values.

It is important to realise that temperatures such as T_{min} and T_{max} are *parameters* of the four-parameter model Eqn. 3.12. These two temperatures represent cut-off points below and above which, respectively, no growth will be observed. They do not necessarily imply that finite growth *will* be observed at temperatures slightly above T_{min} or slightly below T_{max}. In general, it is difficult to obtain accurate data at very low growth rates, since very long measurement times are required. Close to T_{max}, however, growth rates generally change very rapidly for small changes in temperature, so it is possible sometimes to obtain measurable growth to within 1 K of T_{max}, as has been reported for *Moraxella* species strain 4.16 (Ratkowsky *et al.*, 1983).

3.3. ARRHENIUS-TYPE RELATIONSHIPS

In this section we shall progress from models which only consider the apparent activation energy, E_a, of bacterial growth and the influence of temperature on that property, to models which are referred to as 'master reaction models'. These assume that one enzyme in a metabolic sequence is the bottleneck, i.e. is rate limiting, to the bacterial specific growth rate. It is assumed that it is the same enzyme which governs the rate limiting step both above and below the optimum temperature for growth, and that this enzyme is reversibly denatured both by temperatures higher and lower than the optimum. A more rigorous thermodynamic approach is given in Chapter 10, which is foreshadowed by Brandts' model to be described in 3.3.4.

3.3.1. The Arrhenius and Eyring Models

The most simple and straightforward of the Arrhenius-type relationships is that given by Eqn. 3.8, viz.

$$k = A\exp(-E_a/RT)$$

and is intended to fit data only in the suboptimal temperature region. Fitting the model in the form as written above is a problem in nonlinear regression, since the activation energy, E_a, appears in the expression in a nonlinear form. Often, the logarithm of both sides of the equation is taken to yield

$$\ln k = \ln A - E_a/RT \qquad (3.13)$$

A plot of ln k *versus* reciprocal absolute temperature $1/T$ would

Table 3.1 **Cardinal Temperatures for 33 Strains of Bacteria**

Organism	Ref.	T_{min}	T_{opt}	T_{max}
Antarctic strain 755	*a*	251	288	293
Cytophaga johnsonae C21	*b*	265	297-301	307-308
Pseudomonas Group I 16L16	*c*	266	302	310
Pseudomonas Group II 2.3	"	266	300	310
Pseudomonas Group I 6.4	"	267	299	304
Alteromonas CLD38	"	267	299	309
Acinetobacter 6.12	"	268	302	311
Pseudomonas Group II 5.16	"	269	302	313
Flavobacterium (*Cytophaga*) 2.4	"	269	302	310
Flavobacterium (*Cytophaga*) 6.32	"	270	301	310
Acinetobacter 2.55	"	271	301	309
Acinetobacter 4.41	"	271	303	311
Pseudomonas Group I 4.54	"	272	304	310
Pseudomonas aeruginosa	*d*	272	312	320
Moraxella 4.16	*c*	272	303	314
Proteus morganii M68	"	272	310	318
Acinetobacter 3.25	"	274	305	315
Lactobacillus plantarum	e	276	310	318
Escherichia coli	*d*	276	314	322
Vibrio psychroerythrus (ATCC 27364)	"	276	286	293
Vibrio marinus (ATCC 15381)	"	276	288	294
Vibrio marinus (ATCC 15382)	"	277	297	303
Aeromonas 4.29	*c*	277	309	320
Serratia marcescens	*d*	278	308	314
Pseudomonas fluorescens	"	279	312	320
Bacillus megaterium	"	280	314	325
Bacillus subtilis	"	284	312	326
Coccobacillus	*f*	287	326	332
Bacillus coagulans	*f*	290	325	338
Bacillus sp. NCIB 12522	*f*	291	324	335
Bacillus stearothermophilus 238	*c*	303	331	341
Thermus aquaticus	*d*	306	344	357
Bacillus stearothermophilus	"	308	337	349

a: McMeekin (1988); *b*: Reichardt and Morita (1982); *c*: Ratkowsky *et al.* (1983); *d*: Mohr and Krawiec (1980); *e*: Zwietering *et al.* (1991); *f*: Heitzer *et al.* (1991).

yield a straight line if the activation energy E_a and collision factor A were unchanging with temperature. There is overwhelming evidence for biological systems, however, [see, for example, Johnson *et al.* (1974)] that such plots are not linear but curvilinear, the slope becoming increasingly more negative as the reciprocal of temperature (in K^{-1}) increases (see Fig. 3.1). This implies that, instead of a constant activation energy, E_a decreases in a systematic manner with increasing temperature. The goodness of fit of the solid curves in Fig. 3.1 representing Eqn. 3.10 indicates how well the square root model describes the temperature dependence of microbial growth for those data.

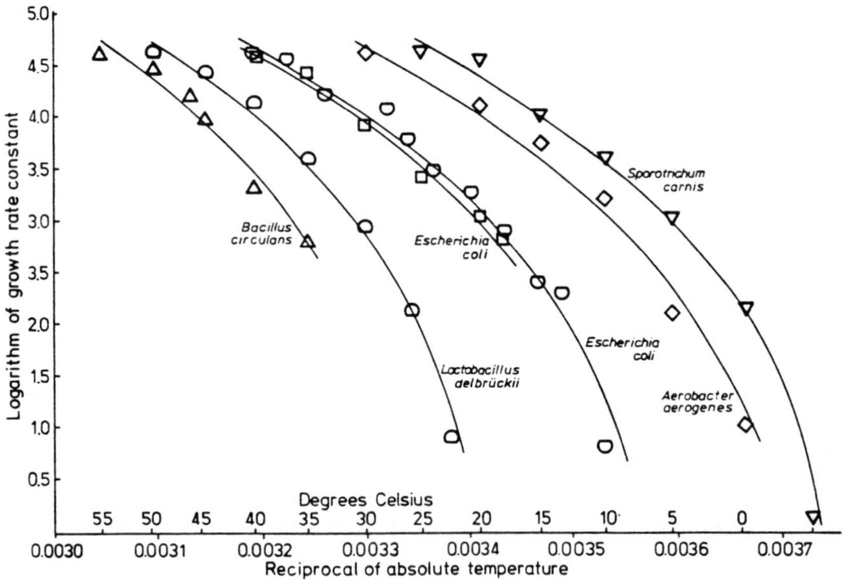

Fig. 3.1 Arrhenius plots for six sets of data redrawn from Johnson *et al.* (1974). The plots show clearly that ln k is not directly proportional to $1/T$, the usual justification for performing an Arrhenius plot. The response is, in fact, curvilinear with the magnitude of the negative slope increasing as the temperature decreases. [From Ratkowsky *et al.* (1982) with permission of the American Society for Microbiology].

An improved model of the Arrhenius-type for the low temperature region is one based on the theory of absolute reaction rates (Eyring, 1935), a simplified form of which may be written as:

$$k = KT\exp(\Delta H^{\ddagger} / RT) \tag{3.14}$$

The major difference between Eqns. 3.14 and 3.8, aside from the theoretical difference between $\Delta H^{\ddagger}$ and E_a, is the presence of an extra temperature factor T preceding the exponential. This is analogous to the collision factor A in Eqn. 3.8 but is temperature dependent. Nevertheless, over the biokinetic range values of $\Delta H^{\ddagger}$ on a reconciliation of Eqns. 3.10 and 3.15 differ by no more than 5% from values of ΔE_{app} given in Table 8.1 (p. 241). The symbol $\Delta H^{\ddagger}$ refers to the enthalpy of activation and K, which in some formulations is a function of the entropy of activation $\Delta S^{\ddagger}$, is a constant, having units of $[\text{time}\cdot\text{temperature(K)}]^{-1}$. Sometimes, Eqn. 3.14 is expressed in logarithmic form as:

$$\ln k = \ln K + \ln T - \Delta H^{\ddagger} / RT \tag{3.15}$$

Unlike the situation with Eqn. 3.13, a plot of ln k *versus* $1/T$ would not yield a straight line if Eqn. 3.15 were valid. Later, we will examine how well Eqn. 3.14 fits actual data sets. It can be said here, however, that the presence of an additional temperature factor in Eqns. 3.14 and 3.15, compared with Eqns. 3.8 and 3.13, makes only a minor difference in the goodness-of-fit of these models to real data. The main reason for preferring models based on absolute reaction rate theory to the basic Arrhenius model is the belief that absolute reaction rate theory has a sounder theoretical basis than the Arrhenius model.

3.3.2. The Johnson and Lewin Model and Variants

The Arrhenius-type relationship most widely used is the model of Schoolfield *et al.* (1981). This is, however, simply a reparameterisation to give better parameter estimation properties, but without change to the model theory, to the model proposed by Sharpe and DeMichele (1977). Both models assume that growth rate is governed by a single rate-limiting enzyme-catalysed reaction, the rate of which may be described by the

Arrhenius equation with additional thermodynamic terms to describe the rate modifying effects of high and low temperature inactivation of the rate-limiting enzyme. Both models are a synthesis of the equations proposed by Hultin (1955) and Johnson and Lewin (1946). Adair *et al.* (1989) reparameterised the model of Schoolfield *et al.* (1981) and expressed it with ln(time), rather than rate, as the dependent variable. The historical development of these models has been described by Heitzer *et al.* (1991), and also by Ratkowsky *et al.* (1991) from which the following is largely drawn.

3.3.2.1. The model of Johnson and Lewin (1946)

$$\text{rate} = \frac{c \cdot T \cdot \exp(-\Delta H^{\ddagger}/RT)}{1 + \exp(\Delta S/R)\cdot\exp(-\Delta H/RT)} \tag{3.16}$$

where:

- c = a constant encompassing: enzyme concentration; substrate concentration; a proportionality constant to account for units of measurement, and the factor kT/h where k = the Boltzmann constant and h = Planck's constant,
- T = temperature in Kelvin
- R = the gas constant
- $\Delta H^{\ddagger}$ = the heat (enthalpy) of activation of the growth rate controlling reaction
- ΔS = entropy of denaturation of the rate controlling enzyme
- ΔH = enthalpy of denaturation of the rate controlling enzyme

In this model the numerator:

$$c \cdot T \cdot \exp(-\Delta H^{\ddagger}/RT)$$

has the form of an Arrhenius relationship and is used to describe the kinetics of the growth-rate-governing reaction, which is assumed to be enzyme catalysed. To account for the deviation of observed growth response from that predicted by the simple Arrhenius model, the model proposes that the growth-rate-governing enzyme becomes inactivated at high temperatures in a manner also described by a unique Arrhenius model:

$$[\text{Enyme}_{\text{active}}] = [\text{Enzyme}_{\text{total}}]\cdot \exp(\Delta S/R)\cdot\exp(-\Delta H/RT)$$

i.e. this function models the temperature dependent transition of the growth-rate-controlling enzyme between an active and inactive state.

Thus, the overall growth rate is determined by the kinetics of the reaction catalysed by the growth-rate-governing enzyme and also the proportion of the population of that enzyme in an active state. These two temperature dependent processes are mathematically 'superimposed' to yield the full model. For temperatures at which the rate of reaction from the active to inactive state is close to zero, the denominator approximates to unity, and the growth rate is predicted to obey the Arrhenius relationship. As the rate of transition to the inactive form of the growth rate-governing enzyme increases [i.e. the term $\exp(\Delta S/R)\cdot\exp(-\Delta H/RT)$ has values greater than zero] the denominator takes values greater than unity. Thus the prediction of the simple Arrhenius equation (i.e. the numerator) is 'corrected' for the effects of enzyme inactivation caused by high temperatures. It should be noted that the Johnson and Lewin (1946) model has been used for either high or low temperature reversible denaturation of enzymes.

3.3.2.2. The model of Hultin (1955)

This model may be expressed:

$$k' = \frac{x \cdot k \cdot T}{h} \cdot \frac{\exp(\Delta S^{\ddagger}/R)\cdot\exp(-\Delta H^{\ddagger}/RT)}{1 + K} \qquad (3.17)$$

where T, R and $\Delta H^{\ddagger}$ have the same meaning as in Eqn. 3.16, and

k' = rate

k = the Boltzmann constant

h = Planck's constant

x = the transmission constant

$\Delta S^{\ddagger}$ = the entropy of the enzyme catalysed reaction

$K = \exp\left[\frac{-\Delta H}{R}\left(\frac{1}{T} - \frac{1}{T_{1/2}}\right)\right]$

$T_{1/2}$ = $\Delta H / \Delta S$; the temperature at which half of the population of the enzyme is active and half has been inactivated by low temperature

ΔH = the enthalpy of (low temperature) denaturation of the enzyme,

ΔS = entropy of denaturation of the rate controlling enzyme

and was developed to describe the effect of temperature on enzyme activity rather than microbial growth rate. Nonetheless, with the parameter K shown in full, this model is mathematically equivalent to the Johnson and Lewin (1946) model. Hultin introduced the term $T_{1/2}$, which could be evaluated by graphical methods to facilitate determination of the parameters ΔH and $\Delta H^{\ddagger}$. This model is analogous to the model of Maier *et al.* (1955).

3.3.2.3. The model of Adair *et al.* (1989)

We will show in 3.3.3.3. that the four parameter model used by Adair *et al.* (1989) is mathematically identical to the Johnson and Lewin model, but for simplicity of discussion we will consider it after presenting the model of Sharpe and DeMichele (1977) and its variants.

3.3.3. The Model of Sharpe and DeMichele and Variants
Although the Eyring equation, Eqn. 3.14 (or 3.15), does not provide a good fit to data in the suboptimal temperature region, an improvement can be made by dividing the right-hand side of Eqn. 3.14 by an exponential expression containing two additional parameters to yield Eqn. 3.16, which was used by Johnson and Lewin (1946) to describe the growth of *E. coli* as a function of temperature. In the example cited, $\Delta H^{\ddagger}$ was about 0.1 of the value of ΔH for the growth of *E. coli* in the presence of quinine. Consequently, the graph of k *versus* T passes through a maximum rate at an optimum temperature, T_{opt}, as discussed earlier in this section in relation to the four-parameter square root model, Eqn. 3.12. Thus, Eqns. 3.12 and 3.16 are direct competitors as candidates for modelling the temperature dependence of the specific growth rate constant k throughout the entire biokinetic temperature range. Both models have four parameters and are equally 'parsimonious'.

It is unlikely that any empirical model can be postulated which could describe real data for k *versus* T with less than four parameters. This is because two parameters are clearly needed to describe the temperature dependence in the low temperature region for T ranging from the minimum temperature for observable growth to temperatures below the optimum growth temperature, and a further two parameters are needed to force the curve to reverse direction, as T exceeds T_{opt}, and plunge rapidly to a zero growth rate. Johnson and Lewin (1946) attribute the marked decline in rate to be in the nature of a reversible protein denaturation, rather than an irreversible, thermal destruction, the latter concept being favoured by many earlier authors cited by Johnson and Lewin. A comparison of Eqns. 3.12 and 3.16 will be made in Chapter 4 using real data.

3.3.3.1. The model of Sharpe and DeMichele (1977)

Although Eqn. 3.16 is capable of fitting growth rate data throughout the entire temperature range for growth, further models have been postulated based upon absolute reaction rate theory, which lead to less parsimonious forms. Sharpe and DeMichele (1977) advanced the six-parameter model (Eqn. 3.18) which may be viewed as an extension of the Johnson and Lewin (1946) formulation, by adding another exponential term to the denominator of the model. Sharpe and DeMichele (1977) intended that their model be applicable to the development of any poikilotherm, including metabolic processes such as growth, respiration or photosynthesis.

The model of Sharpe and DeMichele (1977) is given by the expression:

$$\text{rate} = \frac{T \cdot \exp\left[(\text{Ø} - \Delta H_A^{\ddagger}/T)/R\right]}{1 + \exp\left[(\Delta S_L - \Delta H_L/T)/R\right] + \exp\left[(\Delta S_H - \Delta H_H/T)/R\right]} \tag{3.18}$$

where:

T and R have the same meaning as previously

Ø = a constant, similar to that in the Johnson and Lewin (1946) model encompassing the entropy of activation of the rate controlling enzyme, enzyme concentration, and Planck's and Boltzmann's constants

$\Delta H_A^{\ddagger}$ = the heat (enthalpy) of activation of the growth rate controlling reaction

ΔS_L = entropy of low temperature denaturation of the rate controlling enzyme

ΔH_L = enthalpy of low temperature denaturation of the rate controlling enzyme

ΔS_H = entropy of high temperature denaturation of the rate controlling enzyme

ΔH_H = enthalpy of high temperature denaturation of the rate controlling enzyme

Sharpe and DeMichele's model also assumes a growth-rate-controlling enzyme. The rate of the reaction catalysed by this enzyme, when all of the enzyme population is in the active state, is modelled by the numerator. The denominator includes terms which model the transition between the active and inactive form of the enzyme for both high and low temperature inactivations.

3.3.3.2. The model of Schoolfield *et al.* (1981)

This model is given by the expression:

$$\text{rate} = \frac{\rho\,(25°C)\cdot\frac{T}{298}\cdot\exp\left[\frac{\Delta H_A^{\ddagger}}{R}\left(\frac{1}{298}-\frac{1}{T}\right)\right]}{1+\exp\left[\frac{\Delta H_L}{R}\left(\frac{1}{T_{1/2_L}}-\frac{1}{T}\right)\right]+\exp\left[\frac{\Delta H_H}{R}\left(\frac{1}{T_{1/2_H}}-\frac{1}{T}\right)\right]} \tag{3.19}$$

where all the parameters have the same meaning as those of the Sharpe and DeMichele (1977) model, but with three new parameters defined as follows:

$\rho(25°C) = 298\cdot\exp(\text{Ø} - \Delta H_A^{\ddagger}/298)/R$ and is analogous to Ø in Eqn. 3.18

$T_{1/2_L} = \frac{\Delta H_L}{\Delta S_L}$ and is the temperature at which half of the population of the rate controlling enzyme is active and half has been inactivated by low temperature

$T_{1/2_H} = \frac{\Delta H_H}{\Delta S_H}$ and is the temperature at which half of the population of the rate controlling enzyme is active and half has been inactivated by high temperature

The latter two parameters are analogous to Hultin's $T_{1/2}$ and also represent the low and high temperatures respectively at which the equilibrium constants for the activation/inactivation reaction for the rate controlling enzyme are 1.0.

3.3.3.3. The Model of Adair *et al.* (1989)

Adair *et al.* (1989) developed an alternative form of the Schoolfield *et al.* (1981) model which they claimed had improved numerical stability and increased speed of convergence when data are fitted to the model by the NLIN nonlinear regression procedure of SAS[1]. In this model ln *time* is the dependent variable. The form of the model is:

[1]SAS Institute Inc., SAS Circle, Box 8000, Cary, North Carolina, 27512-800, USA.

$$\ln K = A + (B/T) - \ln T + \ln\,[1 + \exp\,(F + D/T) + \exp\,(G + H/T)] \quad (3.20)$$

where:

K = lag or generation time

A = $\ln 298 - [-\Delta H_A^{\ddagger}/(298 \cdot R) - \ln \rho(25°C)]$

B = $\Delta H_A^{\ddagger}/R$

D = $-\Delta H_L/R$

F = $-\Delta H_L/(T_{1/2_L} \cdot R)$

G = $-\Delta H_H/(T_{1/2_H} \cdot R)$

H = $-\Delta H_H/R$

and where R, T, $\Delta H_A^{\ddagger}$, ΔH_H, ΔH_L, $T_{1/2_H}$, $T_{1/2_L}$ and $\rho(25°C)$ are as defined for Eqns.3.18 and/or 3.19

For data in the suboptimal temperature range the model is simplified by the deletion of the term: $\exp\,(G + H/T)$. If the full expression for $\rho(25°C)$ is substituted into the four parameter form of Eqn. 3.20, it simplifies to the form of the Eqns 3.16 and 3.17.

3.3.3.4. Appraisal of the Sharpe and DeMichele and Schoolfield *et al.* models

Schoolfield *et al.* (1981) recognised that the original formulation of Sharpe and DeMichele (1977) was poorly suited for nonlinear regression, as convergence to the least-squares parameter estimates was difficult to achieve. In addition, the parameter pairs Ø, $\Delta H_A^{\ddagger}$; ΔS_L, ΔH_L; and ΔS_H, ΔH_H, exhibited extremely high correlation resulting in estimates with very high standard errors. They reparameterised Eqn. 3.18 to give Eqn. 3.19.

Although Schoolfield *et al.* (1981) intended that Eqn. 3.19 overcome the problems experienced with nonlinear estimation when the alternative formulation Eqn. 3.18 was used, some difficulty was still experienced in fitting the reparameterised model (Lowry and Ratkowsky, 1983). Both reparameterisations exhibit significant 'intrinsic' and 'parameter effects' curvatures (Bates and Watts, 1980, 1988) and produce substantially biased parameter estimates. A further weakness of both parameterisations is the failure of the models to produce a zero growth rate at any finite value of the temperature T. Instead, the form of Eqns. 3.18 and 3.19, with exponentials in the denominator controlling the shape of the curve at very low and at very high temperatures (see Fig. 3.2), is not particularly suited for modelling processes where the fall off in rate occurs at a faster rate than that given by the exponential of the negative reciprocal of absolute temperature. Decreasing the growth rate constant at extreme temperatures would probably require extra terms in the model to account for irreversible inactivation effects of the enzyme, whereas the present models only incorporate terms that account for reversible inactivation. However, to do so would further increase the complexity of models that already lack parsimony.

3.3.4. Other Arrhenius-type models

3.3.4.1. The Hinshelwood model

Zwietering *et al.* (1991) drew attention to the Hinshelwood model:

$$U_m = k_1 \cdot \exp(-E_1/RT) - k_2 \cdot \exp(-E_2/RT) \tag{3.21}$$

where U_m is the specific growth rate, k_1 and k_2 are frequency factors and E_1 and E_2 are the activation energies of the enzyme reaction and the high temperature denaturation of the reaction *product*, respectively.

In a comparison of the Hinshelwood model with the expanded square root model (Eqn. 3.12), a reparameterisation of the expanded square root model, and the Schoolfield model (Eqn. 3.19), the Hinshelwood model did not model low temperature data well. This will be discussed in Chapter 4.

3.3.4.2. Davey's 'Modified Arrhenius' model

Davey (1989b) proposed a linear Arrhenius model to describe the effects of temperature and water activity on microbial growth rate that will be considered in detail in Chapter 5. When water activity is non-limiting, the water activity term(s) may be removed, reducing the equation to:

$$\ln k = C_0 + \frac{C_1}{T} + \frac{C_2}{T^2} \tag{3.22}$$

where k is rate, T is temperature (K), and C_0, C_1, and C_2 are fitted parameters.

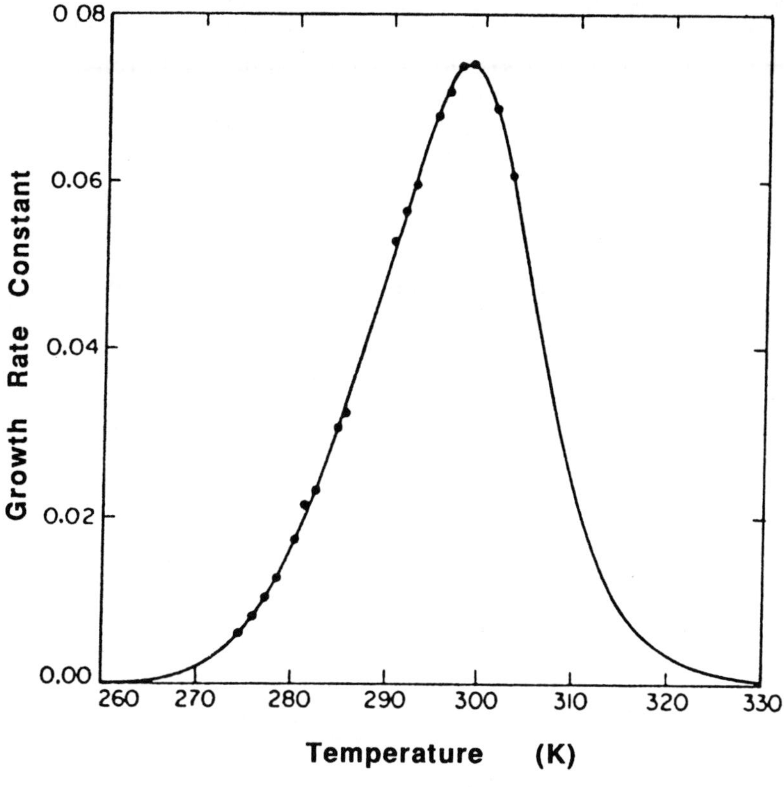

Fig. 3.2 Data for *Alteromonas* sp CLD 38 plotted as rate *versus* temperature (•), and showing the values predicted by fitting a model of the type of Eqn. 3.18 or 3.19 to the data (——). The models obviously fit these data well. The fitted line has been extrapolated deliberately beyond the range of the data to illustrate that by considering only reversible enzyme reactions, the models cannot explain growth throughout the entire temperature range. It is known that bacterial growth is restricted to a finite temperature range (Stanier *et al.*, 1971) yet Eqns. 3.18 and 3.19 predict zero growth rates only at temperatures of + ∞ and - ∞. [Reproduced from Lowry and Ratkowsky (1983), with permission of Academic Press, Inc., Ltd.].

Davey (1991) used Eqn. 3.22 to describe the response of lag phase duration to temperature.

3.3.4.3. The model of Fiolitakis *et al.*

Fiolitakis *et al.* (1986) introduced microbial damage and repair terms to the conventional Arrhenius model. This model was rejected by Heitzer *et al.* (1991) due to the dependence of the model on initial parameter estimates.

3.3.4.4. Brandts' model

Brandts (1967) pointed out that models involving protein denaturation in reversible biological processes were incomplete unless terms involving the thermodynamic consequences of the participation of water molecules in the denaturation reaction were included.

The classical definition of standard Gibbs free energy change in absolute temperature:

$$\Delta G^\circ = \Delta H^\circ - T\Delta S^\circ$$

cannot be used in defining protein denaturation because the equation implies that the enthalpy and entropy terms are temperature independent constants.

The five important structural features which contribute to a difference in the free energy between native and denatured states are shown in Table 10.1 (p. 292). The most important of these in the minimum temperature range for bacterial growth is ΔG_{trans}, the exposure of non-polar side chains of amino acids to water when a globular protein unfolds.

ΔG_{trans} is the sum of the effects of all the amino acid side chains in a protein and can be derived from the empirical expression:

$$\Delta G_{trans} = aT + bT^2 + cT^3$$

where a, b and c are constants based upon values given in Table 10.2.

This (expansion) term is necessary due to the very large change in heat capacity ΔC_p when a globular protein unfolds on low temperature denaturation exposing hydrophobic groups of amino acids to water. The magnitude of this effect may be appreciated from Table 3.2 which shows values of the heat capacity and entropy changes in biochemical reactions at 25°C. The large positive entropy and specific heat changes for protein unfolding are in marked contrast to the small and negative specific heat changes for a normal enzyme.

Table 3.2 **Heat Capacity and Entropy Changes in Biochemical Reactions at 25°C***

System	ΔS_{μ}† cal K^{-1} mol^{-1}	ΔCp cal K^{-1} mol^{-1}
Aldolase + hexitol-1,6-diphosphate	34	-410
GPDH + NAD^+ (pH 8.5)	-45	-750
Lipase + colipase	13.6	-300
Heart LDH + NAD^+	3.5	-84
Heart LDH + NADH	-2.8	-170
Heart LDH-NAD^+ + oxalate	8.0	-340
Heart LDH-NADH + oxamate	-20	-405
Muscle LDH + NADH	9.0	-325
tRNA ligase + isoleucine	19.7	-430
Avidin + biotin	1.3	-240
Haemoglobin + haptoglobin	-73	-940
Rabbit IgG, MOPC 315, MOPC 460 + Dnp-lysine	-7 to -24	-200
Bovine colostrum IgG_2 + Dnp-lysine	8.5	+140
α–Chymotrypsin unfolding (pH 7)	330	+3080
Chymotrypsinogen A unfolding (pH 5, 62°)	440	+3000
Cytochrome *c* unfolding (pH 7)	6.2	+1640
Lysozyme unfolding (pH 7)	140	+1560
Metmyoglobin unfolding (pH 11.5, 63°)	300	+2760
Ribonuclease unfolding (pH 2, 31°)	215	+1220

*ΔC_p appears to be temperature independent in the vicinity of 25°C. All values are average values per site in multisite cases. LDH: lactate dehydrogenase, Dnp: dinitrophenol

†ΔS_{μ}: unitary entropy (see Gurney, 1953).

[*After* Sturtevant (1977) with permission of US National Academy of Science].

Brandts therefore substituted a term for Gibbs free energy in the denominator of the Arrhenius-type Eqn. 3.16:

$$\text{Rate} = \frac{\text{C}\exp(-\Delta E_a/\text{R}T)}{1 + \exp(-\Delta \text{G}°/\text{R}T)} \qquad (3.23)$$

In Chapter 10 we shall show the effect of the magnitude and importance of ΔC_p in determining the shape of rate curves through the biokinetic temperature range. At this juncture it suffices to say that Brandts (1969) pointed out that the magnitude of ΔC_p determines the curvature of an Arrhenius plot. This approach is in marked contrast to the derivations in 3.3.1 to 3.3.3 where tangents to the curve are interpreted as values of constants of the equations.

3.4. RECONCILING THE ARRHENIUS AND BĚLEHRÁDEK EQUATIONS

It will be demonstrated in Chapter 8 that, in order to design and select time-temperature monitors appropriate to the spoilage and shelf life prediction of various foodstuffs, it is useful to be able to interpret data fitted to the Bělehrádek equation in terms of the activation energies described by the Arrhenius approach.

Despite its limitations (see 3.3.1.), the Arrhenius-kinetics paradigm is still widely used. Consequently, McMeekin *et al.* (1988) presented a reconciliation of the Arrhenius Law and the square root model. The derivation is as follows:

The differential form of the Arrhenius Law is:

$$\frac{\text{d} \ln k}{\text{d}\, T} = \frac{\text{E}}{\text{R}T^2}$$

Taking the logarithm of both sides of the square root equation:

$$\sqrt{k} = \text{b}(T - T_{min})$$

and differentiating with respect to temperature yields:

$$\frac{\text{d} \ln k}{\text{d}\, T} = \frac{2}{(T - T_{min})}$$

Hence it follows that the activation energy must be related to temperature by the expression:

$$\Delta E_{app} = \frac{2RT^2}{(T - T_{min})} \tag{3.24}$$

The same approach may be used to reconcile the square root model with Eqn. 3.14. The difference between estimates by the two approaches, however, is small (up to ~5%). Thus, the reconciliation by McMeekin *et al.* (1988) remains a simple, useful tool.

The activation energy, ΔE_{app}, calculated from Eqn. 3.24 is the tangent of the Arrhenius plot of log k *versus* $1/T$ at any temperature T, and changes markedly as a function of temperature. Earlier discussion in this chapter has considered the reasons why the simple Arrhenius equation is likely to be unreliable for determining the 'activation energy' of the modelled response. Accordingly, we choose to denote the value of E_a calculated from the Arrhenius plot, or from Eqn. 3.24, as the *apparent* activation energy and give it the symbol, ΔE_{app}.

Values of ΔE_{app} for typical psychrophilic, psychrotrophic, mesophilic and thermophilic organisms are presented in Table 8.1, in which criteria for the selection of suitable physical and chemical monitors to simulate bacterial spoilage are considered.

Figure 3.3, overleaf, shows square root relationships, fitted to actual data for the growth of bacteria representative of each of the temperature classes of microorganisms, as Arrhenius plots. Using the fitted square root equation, ln(rate) values were generated for 0.00002 K^{-1} increments between T_{min} and T_{opt} for each of the four organisms in Table 1 of McMeekin *et al.* (1988). The parameter 'b' was assigned the value 1.0 to standardise the data. [Note that different 'b' values only move the curve higher or lower on the *ln(rate)* axis of the Arrhenius plot as:

$$\ln\{b^2(T - T_{min})^2\} = \ln b^2 + \ln (T - T_{min})^2$$

i.e. the standardisation of the 'b' values does not alter the shape of the curve, only its 'vertical' position on the Arrhenius plot]. Thus the data sets are 'idealised'.

Using simple linear regression, and starting with the temperature closest to T_{opt}, successive data points were included in the regression

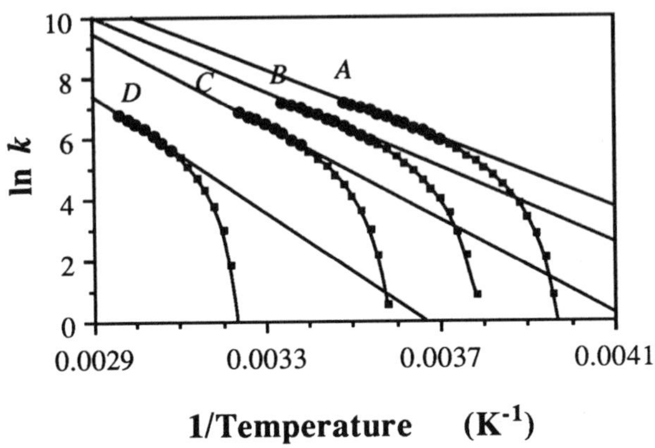

Fig. 3.3 Arrhenius plots of data generated by fitting Bělehrádek models to data for the growth of representatives of each of the four temperature classes of microorganisms (A: psychrophile; B: psychrotroph; C: mesophile; D: thermophile). Growth rates were standardised by setting the value of the Bělehrádek parameter ‘b’ to 1.0 (see text for justification). The plots show that for the higher part of the suboptimal temperature range the Bělehrádek and Arrhenius temperature responses are virtually indistinguishable.

until the correlation coefficients (R^2) fell below the arbitrarily selected value 0.995.

Figure 3.3 shows that even with ‘idealised’ data for temperatures from about 18°C higher than T_{min} and up to nearly T_{opt}, the temperature rate response predicted by the Arrhenius equation is virtually indistinguishable from the Bělehrádek predicted response, except by analysis of the sign of the residuals. Given that real data contains some scatter, it is perhaps evident why some workers have been tempted to fit straight lines to this region and interpret it as a so-called ‘normal region' for growth, i.e. where Arrhenius-type kinetics apply, and to ascribe it special physiological meaning.

We believe that such a delineation is quite arbitrary and that a more objective conclusion is that in the suboptimal temperature region the growth rate response to temperature, when presented as an Arrhenius plot, is a continually downward sloping curve as stated in 3.3.1 and illustrated by Figs. 3.1 and 3.3.

3.5. SUMMARY

In this chapter we have attempted to introduce the reader to the major models that have been proposed to describe the effect of temperature on microbial growth rates. In Chapter 2 we presented a series of criteria that models should meet if they are to be used in this context. The Bĕlehrádek-type models comply in having interpretable parameters such as T_{min} and T_{max} and being parsimonious. Goodness-of-fit of the models to experimental data and parameter stability is the subject of the next chapter in which a comparison is made of several Bĕlehrádek- and Arrhenius-type models in the suboptimal temperature region and in the entire biokinetic range.

4.

COMPARING BĚLEHRÁDEK-TYPE AND ARRHENIUS-TYPE MODELS USING REAL DATA

This chapter examines the equations presented in Chapter 3 and makes a comparison of their efficacy, utilising real data sets, both in the suboptimal and full biokinetic temperature ranges. The importance of making correct stochastic assumptions is stressed, and some guidelines are presented for the selection of models that are likely to be of use in predictive microbiology.

4.1. INTRODUCTION

In Chapter 3 we described two major classes of relationships, *viz.* the Bělehrádek-type and the Arrhenius-type formulations, that have been advanced for modelling the way in which the specific rate constant for bacterial growth changes with temperature. In this chapter, we will examine a variety of data sets and compare the fit of various models to these data. The less variable the observed data, the greater is the chance that one may be able to distinguish between the various models in terms of goodness-of-fit. Thus, it is always a desideratum that the best possible experimental data be acquired when it is a question of trying to compare and distinguish between two or more competing models. We will divide the data sets into two groups, the first being those sets which cover the temperature range from the lowest temperatures at which growth is observable up to just below the optimum temperature for growth. The second group contains those data sets which extend beyond the optimum temperature for growth and span the biokinetic temperature range more fully. We also consider data sets in which the experimental conditions are replicated, enabling one to examine the observed distribution of experimental error and thereby make decisions about what is the appropriate form of the 'stochastic' assumption for use in any subsequent modelling exercise.

4.2. DATA SETS FOR THE SUB-OPTIMAL TEMPERATURE REGION

Examples of good experimental data which have temperatures confined to values below the optimum temperature for growth are to be found in the thesis of Chandler (1988). Some of his data sets pertaining to a single organism, *Pseudomonas* sp. Strain E5.2, isolated from milk, are reproduced in Appendix 4A.1. Data Set 1a involves the time for a specified change in optical density, Data Set 1b is for generation time, and Data Sets 1c, 2 and 3 concern lag times. As illustrated by the straight lines that can be drawn through the data using a 'square root' of rate ordinate and a temperature abscissa (see Figs. 4.1a-c, opposite), the simple square root model, Eqn. 3.10:

$$\sqrt{k} = b(T - T_{min})$$

provides an excellent description of the data. Rate is simply the reciprocal of the response time, whether the time is a specified response time, generation time or lag time. The estimates of the parameters b and T_{min} are given in Table 4.1, along with the correlation coefficients.

The high values of the correlation coefficients, the smallest of which is 0.995, coupled with the visual evidence for three of the data sets shown in Figs. 4.1a-c, where $\sqrt{k}$ of the three growth responses of Data Set 1 are plotted against temperature, tell one that the simple square root model describes the data well. For economy of space, plots of Data Sets 2 and 3 are not shown, but Table 4.1 shows that the fits were equally good.

4.2.1. Comparing Models Using the Sub-optimal Temperature Data Sets of Chandler (1988)

To make a comparison between the various models that have been proposed to fit data such as are presented in Appendix 4A.1, further models, namely the Eyring Model (Eqn 3.15) and the Johnson and Lewin (1946) model, in which the rate was expressed in logarithmic form, as follows:

$$\ln k = \ln K + \ln T - \Delta H^{\ddagger}/RT - \ln\,[1 + \exp(\Delta S/R)\exp(-\Delta H/RT)] \qquad (4.1)$$

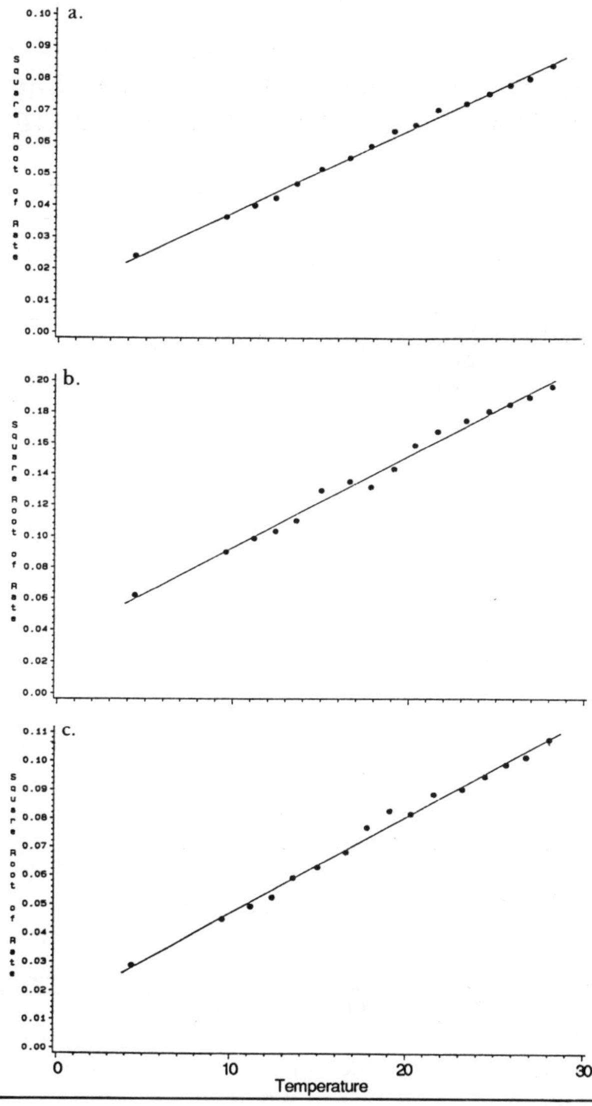

Fig 4.1 Growth of *Pseudomonas* sp. Strain E5.2 modelled with the square root equation (Eqn. 3.10). a) time to ΔOD30; b) generation time; c) lag time. Data sets 1 a-c respectively of Chandler (1988), in Appendix 4A.1. Response time units are minutes, and temperatures shown on the figures are Celsius.

Table 4.1 **Estimated Parameters b and T_{min} and Correlation Coefficients from Square Root Model, Eqn 3.10, for the Data Sets of Chandler (1988) in Appendix 4A.1.**

Data Set		b(x 10^3)	T_{min}	Correlation Coefficient
1a	(ΔOD = 30* Time)	2.600	268.6	0.998
1b	(Generation Time)	5.885	267.4	0.995
1c	(Lag Time)	3.365	269.2	0.996
2	(Lag Time)	4.041	270.0	0.997
3	(Lag Time)	3.292	268.8	0.997

*ΔOD = 30: the time for the optical density of the culture to double from its value at time = 0. (See Fig. 2.5.)

were fitted to the data sets of Chandler (1988).

The fitted models are depicted in Figs. 4.2 a-c, opposite, using the 'neutral' ordinate k, rather than $\sqrt{k}$ or ln k. Recall that rate, k, the reciprocal of time, is transformed in different ways in fitting the three models. Eqn. 3.10 used $\sqrt{k}$, whereas Eqns. 3.15 and 4.1 used ln k. If one were to make comparisons using predicted values of $\sqrt{k}$, Eqn. 3.10 would tend to be unfairly favoured, because least-squares estimation minimises the sum of squares of residuals for that model using a $\sqrt{k}$ scale. If one were to conduct the comparisons using a ln k scale, Eqns. 3.15 and 4.1 would gain an advantage since that scale was used in the least-squares fitting employed with those models. Choosing a k scale is an attempt to neutralise the effect, inherent in least-squares estimation, of inducing homogeneity in the residuals in whatever scale was chosen for the response variable in the minimization process.

It is clear from Figs. 4.2 a-c that the Eyring model shows systematic departures from the observed data, giving fitted values that are too low in the middle of the temperature range, and too high at the ends of the temperature range. The latter high estimates may be attributed to the

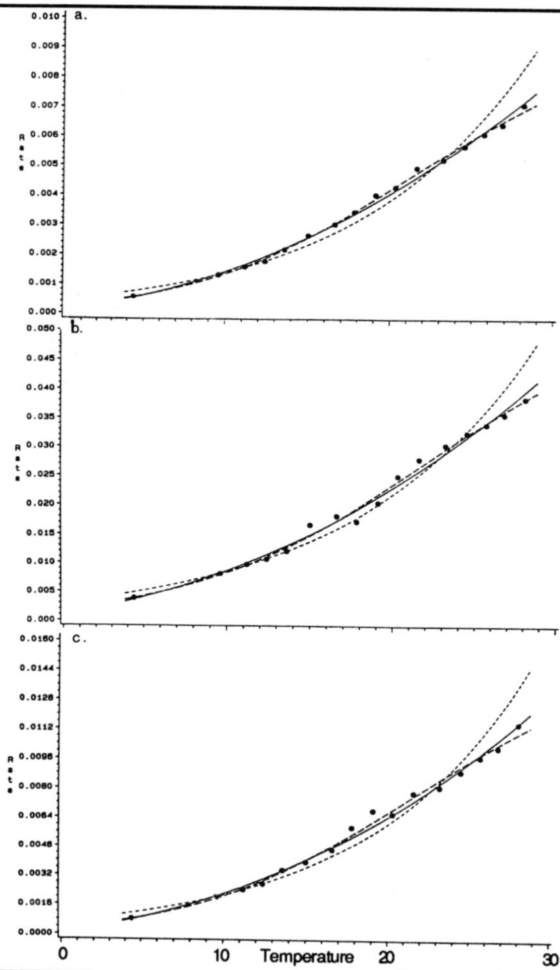

Fig 4.2 Data from Fig. 4.1 a - c for the growth of *Pseudomonas* sp. Strain E5.2 plotted as rate *versus* temperature and modelled with the square root equation (Eqn. 3.10); the Eyring equation (Eqn. 3.15) and the Johnson and Lewin equation (Eqn. 4.1). a) time to ΔOD30; b) generation time; c) lag time. square root equation: —— ; Eyring equation: - - - -; Johnson and Lewin equation: — — . Note that both Eqns. 3.15 and 4.1 were fitted in their *logarithmic* forms although the above graphs depict them as *rate versus* temperature. Data sets 1a-c, respectively, of Chandler (1988), in Appendix 4A.1. Response time units are minutes, and temperatures shown on the figures are Celsius.

lack of enzyme inactivation terms which lead to low estimates in the mid-region as a result of the least squares compensation for the high values. On the other hand, the values of T_{min} in the two parameter square root model were consistent and gave a mean value for the five data sets of 268.8 ± 1.0. Clearly, the Eyring model is not a suitable model for fitting data in the suboptimal temperature portion of the biokinetic temperature range, that is, the temperature range between the lowest temperature at which growth is observed to temperatures just below the optimum temperature for growth.

Also shown in Figs. 4.2 a-c are the fitted curves using the Johnson and Lewin (1946) model. This model provides an excellent fit to the data, the fitted values being very similar to those of the square root model, Eqn. 3.10. Table 4.2 lists the estimates of the parameters ln K, $\Delta H^{\ddagger}/R$, $\Delta S/R$ and $\Delta H/R$ from the fitting of Eqn. 4.1 to the five data sets of Appendix 4A.1.

The estimates of the coefficients for Data Sets 1c, 2 and 3 tend to show considerable variability, although these data sets all relate to the lag phase for the same organism. Thus, although the Johnson and Lewin model provides a good fit to the data in the suboptimal temperature region, there is unlikely to be any physical significance attributable to the estimates themselves in terms of the thermodynamic constants that they supposedly represent. In other words, the non-parsimonious Johnson and Lewin model, having two parameters more than the simple square root

Table 4.2 **Parameter Estimates from Eqn. 4.1 (the Johnson and Lewin Models) to the Data Sets of Chandler (1988).**

Data Set	ln K	$\Delta H^{\ddagger}/R$	$\Delta S/R$	$-\Delta H_L/R$
1a	35.52	13,470	43.45	12,710
1b	30.34	11,500	40.32	11,930
1c	38.79	14,270	45.98	13,430
2	56.75	19,230	53.27	15,160
3	18.13	8,444	92.35	28,040

model, Eqn. 3.10, is best viewed as a 'curve-fitting' model for these suboptimal temperature data, rather than as one which is likely to produce estimates with underlying physical significance.

Appendix 4A.1, in addition to listing the observed ΔOD=30, generation or lag times for the five data sets, also presents the fitted values, converted to time units, resulting from fitting Eqns. 3.10, 3.15 and 4.1 to the data. The results in Appendix 4A.1 confirm the conclusions from visual examination of Figs 4.2 a-c, namely that the Eyring model, Eqn. 3.15, fits the data poorly, whereas both Eqns. 3.10 and 4.1 give excellent fits throughout the whole range of the data.

4.2.2. Comparing Models Using the Sub-optimal Temperature Data Sets of Smith (1985)

We now turn attention to two data sets of Smith (1985), representing the lag time and generation time of coliforms growing in meat. The raw data, given in Appendix 4A.2, range from 10 to 40°C. As with the data of Appendix 4A.1, the simple square root, the Eyring, and the Johnson and Lewin models (Eqns. 3.10, 3.15 and 4.1 respectively), were fitted to these data.

The observed data and the fitted lines for Eqns. 3.10 are shown in Figs. 4.3 a-b (overleaf), in which $\sqrt{k}$, k being the reciprocal of the lag time or the generation time, is plotted against temperature. Straight lines fit the data well, showing that the square root model is a suitable model for these data.

Figs. 4.4 a-b compare the three models using the k *versus* temperature scale and this shows that the Johnson and Lewin model also fits the data very well. As was the case with the data of Chandler (1988) in Appendix 4A.1, only the Eyring model fits poorly, under-predicting the rates in the middle of the low temperature range and over-predicting them at the ends of the range. Table 4.3 gives the estimated T_{min} values from Eqn. 3.10, the two values 276.4 and 275.5 being within one degree of each other.

Table 4.4 gives the parameter estimates from Eqn. 4.1, which differ between the two data sets by a factor of the order of 2. Eqn. 3.24 clearly shows that processes which have the same T_{min} have the same apparent activation energies, E_{app}, at any given temperature below T_{opt}. By inference, one would expect the thermodynamic values of the Arrhenius type models to be very similar for processes with the same T_{min}, particularly $\Delta H^{\ddagger}/R$.

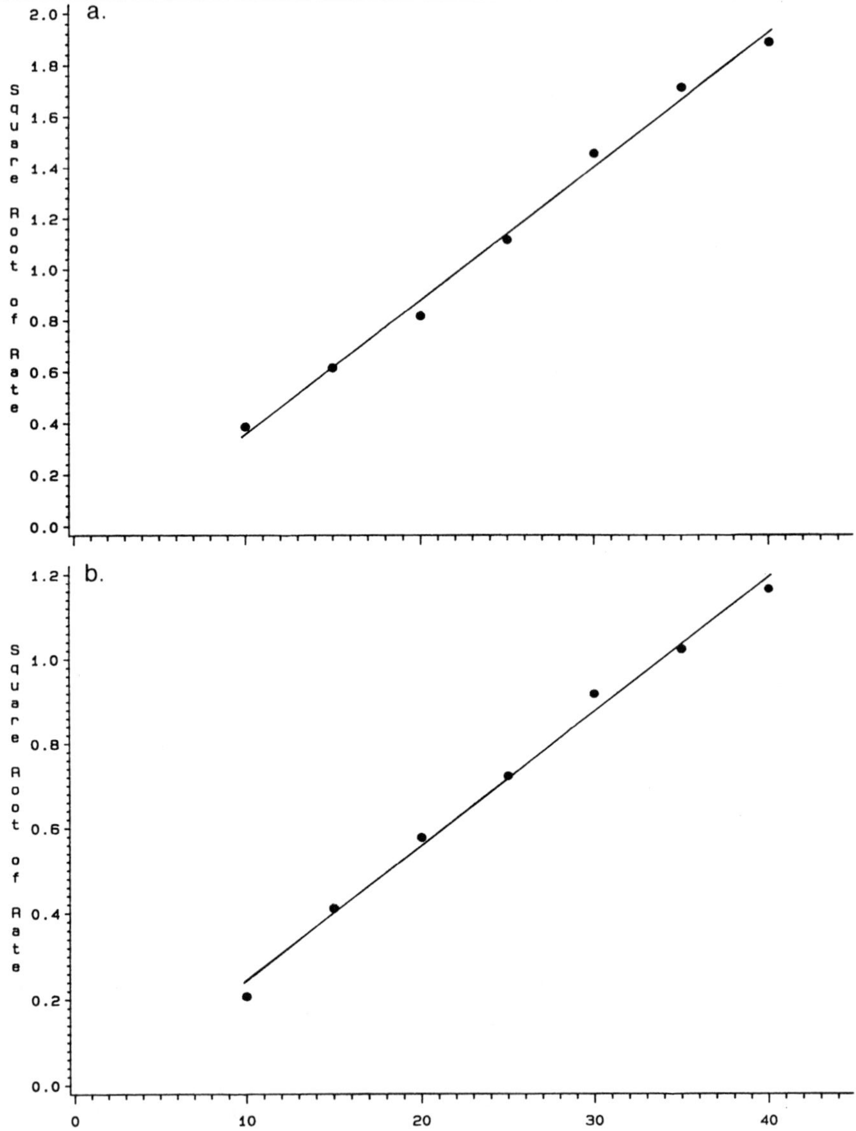

Fig 4.3 Growth of coliforms on meat modelled with the square root equation (Eqn. 3.10). a) generation time (h); b) lag time (h). Data of Smith (1985). Response time units are hours,and temperatures shown on the figures are Celsius.

Table 4.3 **Parameter Estimates for Fitting Eqn. 3.10, the Simple Square Root Model, to the Data of Smith, (1985) in Appendix 4A.2**

Data Set	Response	b (x 10^3)	T_{min}	Correlation Coefficient (R^2)
(a)	Generation Time	52.46	276.4	0.997
(b)	Lag Time	31.59	275.5	0.997

This is borne out by the thermodynamic parameter estimates for Data Sets 1a-c of Chandler which are quite similar for the three responses measured (Table 4.4). For Data Sets 2 and 3, however, the parameter estimates are not consistent with the previous data sets, even though the responses are for the same organism, and as we have shown, for the same T_{min}. Similarly, for the data of Smith, quite different parameter estimates are given for the lag and generation time data (Table 4.4) even though T_{min} is constant. That estimates of these thermodynamic parameters are not constant as predicted by Eqn. 3.24 reinforces the conclusion that the Johnson and Lewin model is a ‘curve-fitting’ model rather than one whose parameter estimates can be viewed as having physical significance.

The fitted values from the three models using the ‘time’ scale are given in Appendix 4A.2, which shows that the square root and Johnson and Lewin models give similar predictions at temperatures between 25 and 40°C, but that there are differences at lower temperatures, with the fitted value from the Johnson and Lewin model being closer to the observed generation or lag time at 10°C, the lowest temperature in the data set. This tendency of the Johnson and Lewin model to predict closer to the observed data value at the lowest observed temperature is also evident in the results in Appendix 4A.1 and is a consequence of two main factors, one of which is that Eqn. 4.1 has two more parameters than Eqn. 3.10. The other factors will be discussed in detail in 4.4.

Table 4.4 **Parameter Estimates from Fitting Eqn. 4.1, the Johnson and Lewin Model, to the Data Sets of Smith (1985) in Appendix 4A.2.**

Data Set	Response	ln K	$\Delta H^{\ddagger}/R$	$\Delta S/R$	$-\Delta H_L/R$
1	Generation Time	40.31	13,530	43.17	13,100
2	Lag Time	101.60	31,190	89.84	25,810

4.2.3. Conclusions Regarding Sub-optimal Temperature Modelling

It is clear from the results depicted in Figs. 4.2 a-c and 4.4 a-b, that the Eyring model (Eqn. 3.15) does not provide a good fit to observed data and is therefore not a suitable model for predictive microbiology. It will not be considered further.

The simplest model of those considered is the two-parameter square root model (Eqn. 3.10). Experience has shown that this is an excellent expression for modelling the temperature dependence of bacterial growth provided that the data are confined to the region between the lowest temperature for which growth is observed and a temperature just below the optimum temperature for growth, that is, the temperature above which the relationship between $\sqrt{k}$ and temperature starts to become curvilinear. Typical examples of how well the square root model fits real data are given in Figs. 4.1 a-c and 4.3 a-b. Eqn. 3.10 is very easy to fit as a linear regression model in the form:

$$\sqrt{k} = c + bt \qquad (4.2)$$

where t is temperature in degrees Celsius.

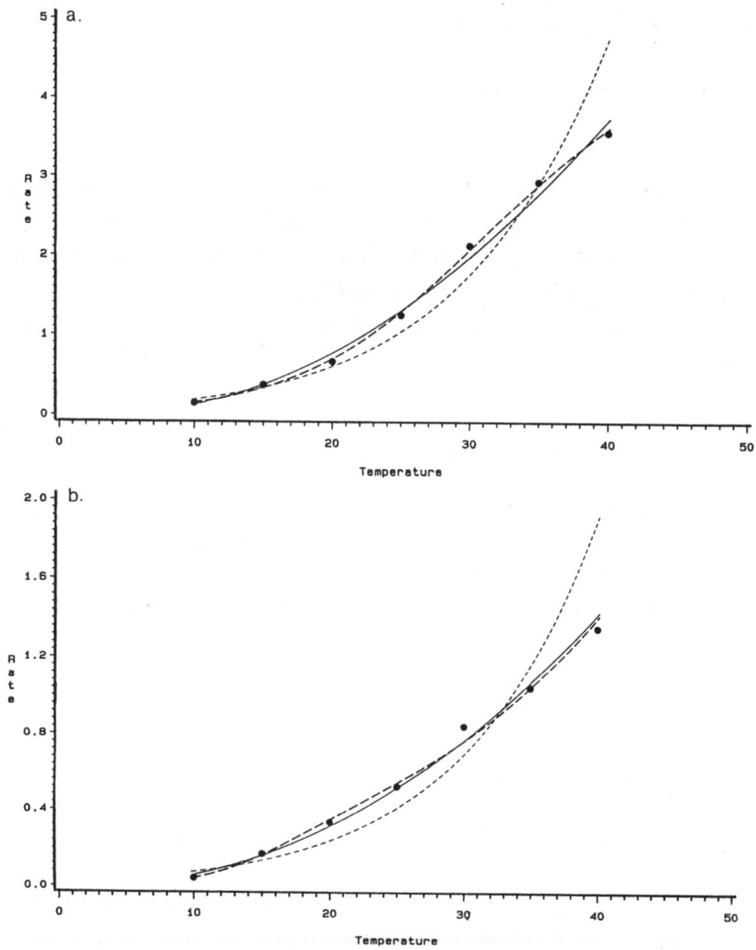

Fig 4.4 Data (• • •) of Fig. 4.3 a and b for the growth of coliforms on meat plotted as rate *versus* temperature and modelled with the square root equation (Eqn. 3.10,—); the Eyring equation (Eqn. 3.15,-----); and the Johnson and Lewin equation (Eqn. 4.1,– – -). a) generation time (h); b) lag time (h). Note that, as with Fig.4.2 a - c, the data were fitted to the logarithmic forms of the models, Eqns. 3.15 and 4.1. Data of Smith (1985). Response time units are minutes and temperatures shown on the figures are Celsius.

This can be fitted using any simple linear regression program to obtain estimates of the parameters b and c. Eqn. 4.2 can then be converted into the form of Eqn. 3.10, which contains the cardinal temperature T_{min}, by using the transformation:

$$T_{min}\ (\mathrm{K}) = 273.16 - (c/b).$$

4.3. SELECTING SUITABLE MODELS FOR PREDICTIVE MICROBIOLOGY FOR THE FULL BIOKINETIC TEMPERATURE RANGE USING THE DATA OF ZWIETERING *ET AL.* (1991)

We now consider data sets that span the temperature range which includes the optimum temperature. For this purpose we will first consider the excellent data set of Zwietering *et al.* (1991) who studied the growth of *Lactobacillus plantarum* at 18 temperatures from 6-43°C.

Growth curves were fitted by Zwietering *et al.* (1991) to the 'modified Gompertz' function (Eqn. 2.11) to give estimates of the specific growth rate and lag phase duration. These estimates were used to select the best model from several describing the correct general response to temperature. For specific growth rate, the competing models were the extended square root model (Eqn. 3.12); a modification of Eqn. 3.12 due to Zwietering *et al.* (1991) in which the decline of growth rate at superoptimal temperatures is described by an exponential function and not by the square of an exponential function, i.e:

$$k = [b(T - T_{min})^2 \{1 - \exp[c(T - T_{max})]\} \; ; \qquad (4.3)$$

the Schoolfield model (Eqn. 3.19); and the Hinshelwood model (Eqn. 3.21).

Figs. 4.5 a-d illustrate that each of the models describes the correct general response and selection was based on the statistical criteria given in 2.6.

The Hinshelwood model does not fit the data well at low temperatures (Fig. 4.5a) due to the lack of a cold denaturation term in the equation. Further, the parameters were highly correlated making their values difficult to estimate and having wide confidence limits. Despite the model being based on Arrhenius kinetics, the activation energies estimates fell well outside the range expected for enzyme catalysed reactions and heat denaturation. For these reasons, Zwietering *et al.* (1991) concluded that the Hinshelwood model was inappropriate.

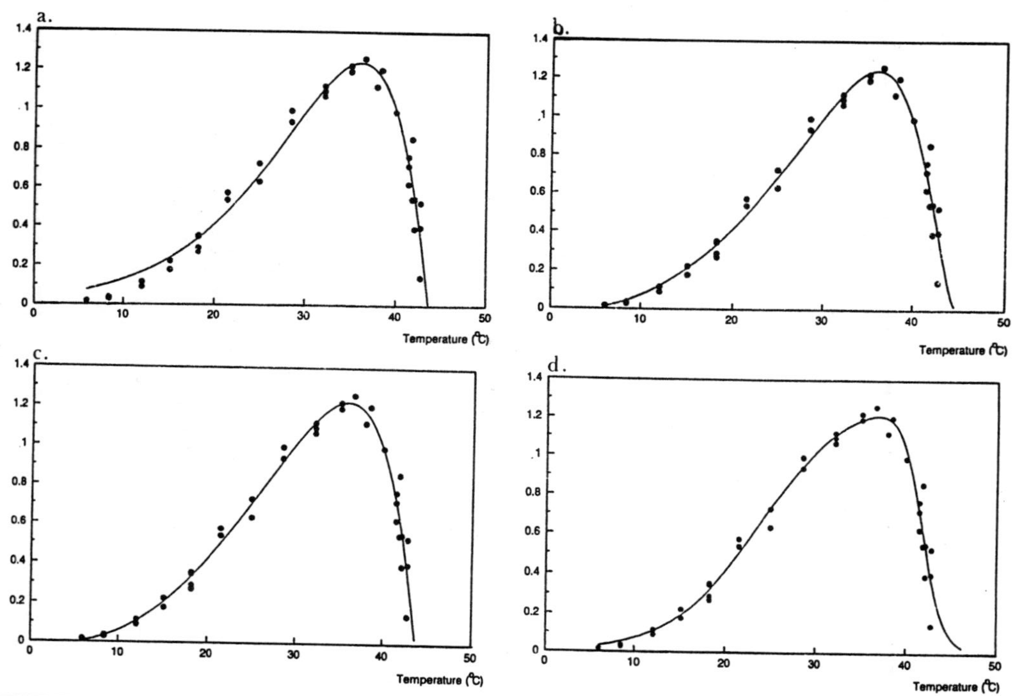

Fig 4.5. Growth rates of *Lactobacillus plantarum* modelled with: a) the Hinshelwood model (Eqn. 3.21); b) the square-root model (Eqn. 3.12); c) the modified square-root model (Eqn. 4.3); and d) the Schoolfield model (Eqn. 3.19). Symbols: • estimated growth rates; —— the model. [Reproduced from Zwietering *et al.* (1991) with permission of the American Society for Microbiology].

The square root models and the Schoolfield model (see Figs. 4.5 b-c and Fig. 4.5d respectively) fitted the data well, but the latter was rejected by Zwietering *et al.* (1991) because it has two additional parameters and the estimates of the parameter representing enthalpy of activation were negative. The extended square root models (Eqns. 3.12 and 4.3) were chosen and the latter preferred as it had a slightly lower residual mean square total than the former.

When considering the temperature dependence of lag phase duration, Zwietering *et al.* (1991) transformed the data logarithmically and fitted the transformed response to the logarithm of the inverse of the square root model (Eqn. 3.12), to give:

$$\ln\ (\text{lag time}) = \ln\ [b(T - T_{min})\{1 - \exp[c(T - T_{max})]\}]^{-2} \qquad (4.4)$$

They also used a hyperbolic equation:

$$\ln\ (\text{lag time}) = \frac{p}{(T - q)} \qquad (4.5)$$

where q is the temperature at which the lag time is infinite and p is a measure of the decrease in lag time when the temperature is increased.

Zwietering *et al.* (1991) obtained reasonable fits to the data using the inverse of the two-parameter square root model, Eqn. 3.10, (that is rearranged so that *time* was the dependent variable) and the two-parameter hyperbola model (Eqn. 4.5). The latter, having a lower residual mean square error, was preferred by those authors. As the data set included sub- and super-optimal temperatures, one might have expected that a model with a greater number of parameters, such as Eqn. 4.4 or the inverse of Eqn. 4.3 expressed as follows:

$$\ln\ (\text{lag time}) = \ln\ [b^{-1}(T - T_{min})^{-2}\{1 - \exp[c(T - T_{max})]\}^{-1}] \qquad (4.6)$$

might be required. Zwietering *et al.* (1991) did in fact fit these more complicated equations and found that they gave somewhat smaller sums of squares than the hyperbola model (Table 4.5). These models might be preferred in situations where the lag times in the optimum to maximum region of growth increase markedly.

Table 4.5 **Some Models Considered by Zwietering *et al.* (1991) and Some Statistics Comparing Their Performances on *L. plantarum* Data Set (38 points) for Describing the Lag Time**

Model	No. of parameters	Error degrees of freedom, df	RSS_2 (Error sum of squares)	f (Test statistic for lack of fit)
Eqn. 4.5	2	36	9.71	0.357
Eqn. 3.10	2	36	18.3	1.60
Eqn. 4.4	4	34	9.14	0.238
Eqn. 4.6	4	34	9.21	0.221

Adair *et al.* (1989) have argued that the simple square root model does not fit data in the low temperature region as well as does the Johnson and Lewin model (which they referred to as the "Schoolfield model"). The basis of their argument was a comparison of lag or generation times after (i) fitting Eqn. 3.10 (or 4.2) in the form $\sqrt{k}$ versus temperature and then converting the fitted values $\sqrt{k}$ to predicted times by squaring the values of $\sqrt{k}$ and taking the reciprocal of the results, (ii) fitting Eqn. 3.20, with the term 'exp(G + H/T)' deleted, as $\ln k$ versus temperature and then converting the fitted values of $\ln k$ to predicted times by exponentiating the values of $\ln k$ and taking the reciprocal of the results. With this approach, Adair *et al.* (1989) showed that the predicted times based on the use of Eqn. 3.20 agreed more closely with the observed lag or generation times than did the predicted times based on Eqn. 3.10.

Their contention appears to be borne out by the results in Appendix 4A.1 and 4A.2, where predicted times are calculated using their approach of converting fitted rates into predicted times, using data sets with maximum temperatures below the optimum temperature for growth. It is seen that the Johnson and Lewin model tends to give predictions that agree with the observed values better than the predictions obtained using the square root model, the discrepancy being most pronounced at the lowest observed temperature, where the rate of bacterial growth is smallest and hence the lag or generation time is greatest.

Further information on the comparison of the performance between competing models is provided by data on the lag phase duration of air-packed beef mince, due to Adair *et al.* (1989), reported in Appendix 4A.3. Those authors used Mean Squared Error (MSE) to compare the models, a criterion appropriate only if lag times are normally distributed. The comparison in Appendix 4A.3, however, is made using Pearson residuals, which will be described in Section 4.4., coupled with the more realistic assumption that the distribution of lag times is 'inverse Gaussian'. The Pearson residuals in the table in Appendix 4A.3 do support the contention that the four-parameter form of Eqn. 3.20 gives predicted values closer to the observed data values in the low temperature region than does the two-parameter square root model, Eqn. 3.10. The overall Pearson goodness-of-fit statistic, which is simply the sum of squares of the Pearson residuals in the table, is lower for Eqn. 3.20 than for Eqn. 3.10. A similar conclusion is obtained if one uses another measure of goodness-of-fit, the 'deviance' measure, also to be described in Section 4.4. Similar conclusions are obtained if one assumes that the error distribution of lag times follows another long-tailed distribution, the 'gamma' distribution. The success of the Schoolfield/Johnson and Lewin models, *vis à vis* the square root model, is achieved at the expense of parsimony, the former models requiring two extra parameters. In addition, a point overlooked in the debate which ensued the Adair *et al.* (1989) paper (see McMeekin *et al.*, 1989; Davey, 1989b; Kilsby, 1989; Ratkowsky *et al.*, 1991), is the fact that lag times and generation times in the low temperature region, where growth is slow, are extremely variable; the range of the observed responses is greatly in excess of any discrepancy between their mean value and the predicted values from *any* of the models. Therefore, the question of which model more closely fits an observed time response in the low temperature region is more of academic than practical interest.

4.4. THE IMPORTANCE OF STOCHASTIC ASSUMPTIONS

As discussed in Chapter 2 and by Ratkowsky *et al.* (1991) least-squares modelling, in common with modelling with various other criteria, involves consideration of both the 'deterministic' part of the model, that is, the relationship between the response (dependent) variable and the explanatory (regressor or independent) variable, and the 'stochastic' part of the model, that is, the error assumption. We will try to show here that it is of paramount importance when comparing the efficacy of models to use a consistent assumption for *all* models that are being fitted to a given set of data. One can operate directly in the time domain using a regression procedure called 'generalised linear regression modelling', which guarantees unbiasedness if one's intention is to work directly with time as the response variable. Alternatively (see 2.4.1.2.), one can utilise an older technique employing 'variance-stabilising transformations'. Both approaches lead to similar answers, particularly if one makes a bias correction for the predicted values when using the latter method. The advantage of the former technique is that there is no correction for bias as one works directly with time, the response of interest. Decision-making tends to be more complex, however, because the probability distributions (e.g. 'gamma' and 'inverse Gaussian') of a time-based random variable are less familiar to most users than a normally distributed random variable. Additionally, use of a transformation exploits the greater development of statistical procedures based upon the assumption of normality, thereby simplifying the approach to such questions as which terms included in the regression model are significant.

4.4.1. Generalised Linear Regression Modelling

The important paper of Nelder and Wedderburn (1972) extended the technique of regression modelling using a normal (Gaussian) error assumption to other members of the 'exponential' family of distributions. These include the Poisson, binomial, multinomial, gamma and inverse Gaussian distributions. The binomial distribution also has implications for predictive microbiology, especially where 'probability modelling' is being undertaken, but otherwise attention will be confined to the normal, Poisson, gamma and inverse Gaussian (the latter also being known as 'inverse normal') distributions. These distributions can be described by a 'variance/mean' relationship given by the simple expression:

$$V = c\mu^n \tag{4.7}$$

where:

- μ is the mean of the probability distribution
- V is the variance of the probability distribution
- n is an integer exponent having values 0, 1, 2, or 3, and
- c is a constant

The four distributions of the exponential family which may be represented by Eqn. 4.7 are summarised in Table 4.6. The variance, V, of the distribution is increasingly more dependent upon the mean, μ, as the exponent, n, increases from 0 to 3. For the normal distribution there is no variance/mean dependence but, with the other distributions, the extent to which the variance depends upon the mean is determined by the value of the exponent, n. In the classical or 'standard' Poisson process, such as occurs with radioactive decay, time between telephone calls on a busy trunk line, and in various 'queueing' or waiting problems, the variance equals the mean, but more generally there is a scale parameter, c, which may differ from 1.0. For the gamma distribution, the variance depends upon the square of the mean (alternatively, the standard deviation is proportional to μ), but the distribution also has a positive scale parameter c. The least familiar of the distributions is the inverse Gaussian, which has the variance proportional to the cube of the mean, with an arbitrary positive scale parameter c.

Generalised linear regression (GLR) modelling [see McCullagh and Nelder (1989) for a monograph on the subject] enables one to fit any of the above distributions to data sets. Two available computing packages for carrying out this modelling process are GLIM (1978) and GENSTAT (1987), marketed in the United Kingdom and the United States by the Numerical Algorithms Group, Ltd. These packages perform an analysis of data having the distributions given in Table 4.6, and utilise a measure of goodness-of-fit known as the 'deviance'. Residuals are correspondingly called 'deviance residuals'. There is more than one kind of residual that

Table 4.6 **The Distributions Encompassed by Equation 4.7, for $0 \leq n \leq 3$**

Distribution	Form of Distribution	n	c
Normal	$V = c$	0	$c \geq 0$
Poisson	$V = c\mu$	1	$c = 1$ gives the standard Poisson case
Gamma	$V = c\mu^2$	2	$c > 0$
Inverse Gaussian	$V = c\mu^3$	3	$c > 0$

can reasonably be defined, however, and correspondingly, more than one definition of goodness-of-fit. McCullagh and Nelder (1989) consider two other kinds of residuals, the Anscombe residuals and Pearson residuals, and the goodness-of-fit measures based upon them. The Anscombe residuals are based upon finding transformations that make the transformed variable as close to normal as possible. They are used mainly as a diagnostic tool for checking model assumptions.

The Pearson residual is just the simple residual $y - \hat{\mu}$ scaled by the estimated standard deviation of the response variable y. The name is derived from the fact that, for the Poisson distribution, the Pearson residual is simply the square root, with the appropriate sign, of the contribution which that observation makes to the Pearson X^2 goodness-of-fit statistic. This definition of a residual makes it clear that the difference between the observed and fitted values needs to be scaled by a function of the estimated mean, $\hat{\mu}$. The goodness-of-fit measures are simply the sums of squares of these quantities, and Table 4.7 summarises the deviance and Pearson measures of goodness-of-fit for various distributions. For the Normal distribution, the deviance is just the residual sum of squares, and when averaged over the N data points becomes the mean square error, which was used by Adair *et al.* (1989). Such a measure of goodness-of-fit can be appropriate only to the Normal

Table 4.7 **Two Measures of Goodness-of-fit in Use in Generalised Linear Modelling**

Distribution	Deviance	Pearson
Normal	$\Sigma(y - \hat{\mu})^2$	$\Sigma(y - \hat{\mu})^2$
Poisson	$2\Sigma[y \ln(y / \hat{\mu}) - (y - \hat{\mu})]$	$\Sigma(y - \hat{\mu})^2 / \hat{\mu}$
gamma	$2\Sigma[-\ln(y / \hat{\mu}) + (y - \hat{\mu}) / \hat{\mu}]$	$\Sigma(y - \hat{\mu})^2 / (\hat{\mu})^2$
inverse Gaussian	$\Sigma(y - \hat{\mu})^2 / \hat{\mu}^2 y$	$\Sigma(y - \hat{\mu})^2 / (\hat{\mu})^3$

error assumption; even so, most workers would divide by the degrees of freedom, N - p, where p is the number of parameters of the model, rather than by N, and use the resulting residual variance rather than the mean square error. For the other distributions, in which the mean and variance are related, there is a term involving the mean fitted response in the denominator of the deviance expression. In comparing goodness-of-fit of different models, one cannot simply average the difference between observed and fitted responses over the whole of the experimental conditions. The formulae in Table 4.7 tell one that the averaging process is a more complicated one, involving a 'scale' factor depending upon the size of the mean response at each condition.

To illustrate how to decide which, if any, of the distributions given in Table 4.6 are suitable for data that occur in predictive microbiology, we consider the data of Neumeyer (1992) presented in Appendix 4A.4. These data, for generation time for *Staphylococcus aureus* 3b, were obtained courtesy of K. Neumeyer and T.A. McMeekin of the University of Tasmania. For each of the six temperatures in Appendix 4A.4 for which there are replicated data, estimates of the sample mean, $\hat{\mu}$, the sample variance, $\hat{V}$, and of the ratios $\hat{V}/\hat{\mu}$, $\hat{V}/\hat{\mu}^2$, $\hat{V}/\hat{\mu}^3$, which aid in deciding which of the distributions is a suitable error description for these data, are given in Table 4.8. The ratios $\hat{V}/\hat{\mu}$, $\hat{V}/\hat{\mu}^2$, $\hat{V}/\hat{\mu}^3$ are plotted in Fig. 4.6, overleaf, and indicate that the inverse Gaussian assumption seems to be the best of the three assumptions for these data, since the ratio $\hat{V}/\hat{\mu}^3$ shows the least dependence upon $\hat{\mu}$.

We choose to compare the performance of three different competing models on the data set in Appendix 4A.4. The three models are the square root model (Eqn. 3.10), but written with rate, k, being replaced by the reciprocal of the generation time θ:

$$\sqrt{1/\theta} = a + bT \qquad (4.8)$$

a 'response surface' model, such as is used by Little *et al.* (1992), but for only one explanatory variable, temperature:

$$\ln \theta = a + bT + cT^2 \qquad (4.9)$$

and the third a 'Davey' model (*cf.* Davey *et al.*, 1978):

Table 4.8 **Estimates $\hat{\mu}$ of the Mean μ and $\hat{V}$ of the Variance V, and of the Ratios $\hat{V}/\hat{\mu}$, $\hat{V}/\hat{\mu}^2$, $\hat{V}/\hat{\mu}^3$, for the Data of Neumeyer (1992). (See Appendix 4A.4).**

Temperature (°C)	$\hat{\mu}$	$\hat{V}$	$\hat{V}/\hat{\mu}$	$\hat{V}/\hat{\mu}^2$	$\hat{V}/\hat{\mu}^3$
12.5	638.7	100 900.2	157.99	0.247	0.000387
15.0	354.2	3 289.2	9.29	0.026	0.000074
20.0	189.6	5 273.3	27.82	0.147	0.000774
28.0	42.38	21.53	0.51	0.012	0.000283
30.0	41.99	29.85	0.71	0.017	0.000403
35.0	30.15	21.82	0.72	0.024	0.000796

$$\ln \theta = a + b/T + c/T^2 \tag{4.10}$$

These models all possess a basic structure which is required by the general form of the generalised linear regression models. They have an expression which is linear in its coefficients, referred to as a 'linear predictor', on the right-hand side of the equation which is related to the response variable θ by means of a 'link function'. For the square root model, the form of the expression on the left-hand side of Eqn. 4.8 is a 'power law', with an exponent of -0.5, whereas for the response surface and Davey models, the link function is 'logarithmic'. Link functions such as these are available in GLIM (1978) or GENSTAT (1987) for fitting generalised linear regression models. It is important to realise that despite the different appearance of the left-hand sides of Eqns. 4.8 - 4.10, it is the same error distribution for θ which applies for each model. Hence, if an inverse Gaussian distribution is assumed for θ, all these models will fit θ using that distribution; similarly, if gamma error is assumed, each of the equations will fit a gamma distribution for θ. Thus, comparisons of the various models, such as whether one of these fits the data better than the others, can be performed because the same response variable θ applies to all the equations.

Fig. 4.6 (Opposite) Plots of the ratios $\hat{V}/\hat{\mu}$, $\hat{V}/\hat{\mu}^2$, $\hat{V}/\hat{\mu}^3$ for the data of Neumeyer (1992). Visual appraisal and simple linear regression of the data suggest that the ratio $\hat{V}/\hat{\mu}^3$ is least dependent upon the magnitude of the mean generation time and, in consequence, that the inverse Gaussian error assumption is most appropriate to these data.

Table 4.9 presents some results comparing the fitting of Eqns. 4.8 - 4.10 on the data sets of Neumeyer (1992). From this table it is seen that there is very little difference between the performance of the three models; the response surface and Davey models fit the data almost identically. The square root model fits somewhat less well, but has one less parameter.

A further test of whether the models fare equally well can be made by examining predicted values. Table 4.10 gives predicted values for all three models using the temperatures actually observed in Neumeyer's data set. The table lists the range of observed generation times at each temperature, and the predictions from the response surface model, Eqn. 4.9, the Davey model, Eqn. 4.10, and the square root model using three different methods, *viz*:

(i) the 'usual' method of fitting $\sqrt{1/\theta}$ vs. T by linear regression, which involves direct use of Eqn. 4.8,
(ii) use of generalised linear regression (GLR) assuming inverse Gaussian error, and
(iii) use of GLR assuming gamma error.

A similar exercise was carried out using the data of Chandler (1988) in Appendix 4A.1 and the data of Smith (1985) in Appendix 4A.2, but confined only to the lowest observed temperatures. The results of fitting the various models are presented in Table 4.11. Perusal of entries in Tables 4.10 and 4.11 does not reveal a tendency for any one model to either over-predict or under-predict responses at any temperature, including the low temperatures at which arresting bacterial growth is of importance in food microbiology.

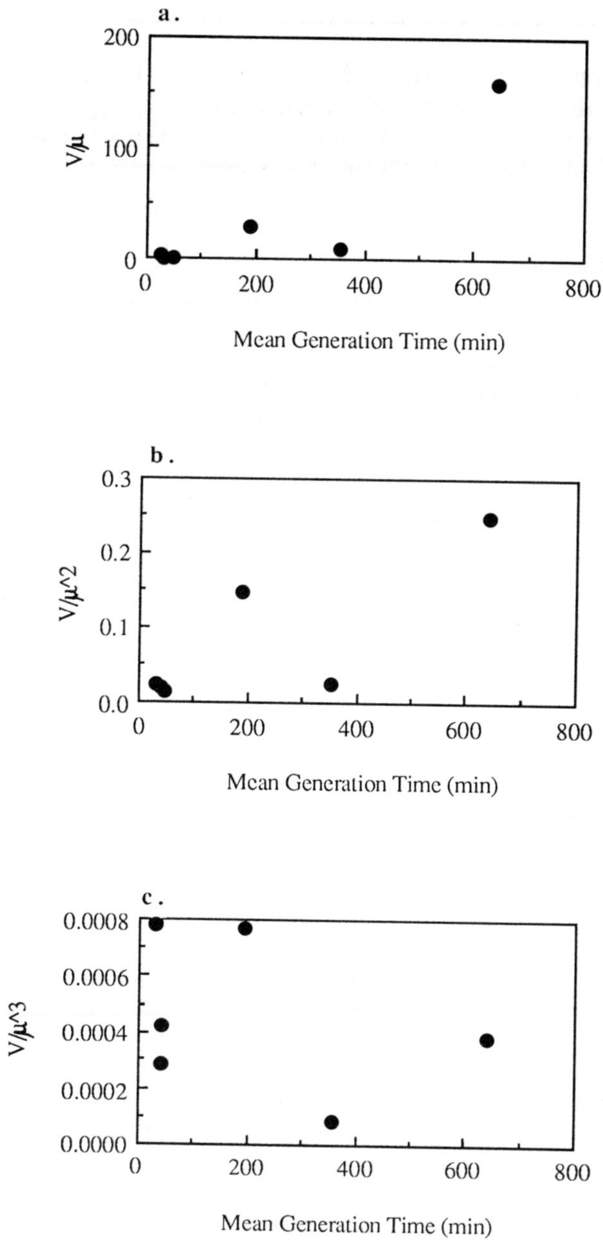
a.
200
100
0
V/μ
0
200
400
600
800
Mean Generation Time (min)
b.
0.3
0.2
0.1
0.0
V/μ^2
0
200
400
600
800
Mean Generation Time (min)
c.
0.0008
0.0006
0.0004
0.0002
0.0000
V/μ^3
0
200
400
600
800
Mean Generation Time (min)

Table 4.9 **Performance of the Models Given by Eqns 4.8 - 4.10 on the Data of Neumeyer (1992), Using the Deviance as a Measure of Goodness-of-fit, Assuming Inverse Gaussian Error.**

Eqn. 4.8 Square root model

	df	deviance	mean deviance	deviance ratio
Regression	1	0.65121	0.65121	739.06
Residual	53	0.04670	0.00088	
Total	54	0.69791	0.01292	

Eqn. 4.9 Response Surface Model

	df	deviance	mean deviance	deviance ratio
Regression	2	0.66396	0.33198	508.47
Residual	52	0.03395	0.00065	
Total	54	0.69791	0.01292	

Eqn. 4.10 Davey Model

	df	deviance	mean deviance	deviance ratio
Regression	2	0.66347	0.33174	500.90
Residual	52	0.03444	0.00066	
Total	54	0.69791	0.01292	

4.4.2. Variance-stabilising Transformations

An alternative approach to the use of generalised linear models is to seek a transformation of the response variable that makes the error distribution close to that of a normal distribution. Such transformations have the property of simultaneously 'stabilising' the variance and tending to make regression models and analysis of variance models more 'additive'. Some variance-stabilising transformations have been known for some time in statistics and biometrical applications, and the method for obtaining them has been described [e.g in Snedecor and Cochran (1967) p. 325]. Their rationale is based upon the use of the first-order term of a Taylor-series expansion for the variance of a function of a random variable. For the four probability distributions considered in this chapter, this leads directly to the variance-stabilising transformations given in Table 4.12.

For normally distributed data one uses the untransformed mean directly. For the Poisson distribution the resulting 'square root' transformation is well established, and the approximate correction for the bias in predicted mean is also well known (see Snedecor and Cochran, 1967, p. 327). For the gamma distribution (or, more generally, for any data set for which it appears that the variance is proportional to the square of the mean), taking the logarithm of y to stabilise the variance is also well known; for example, Snedecor and Cochran (1967, pp. 329-330) give an example of its use. Predicted values of y (that is, on the original untransformed scale of measurement) obtained simply by taking the anti-logarithm of the fitted transformed values of y will be biased (Finney, 1941), and a correction for the bias is needed. The formula for

Table 4.10 **Prediction of Response Times by Eqns. 4.8 - 4.10 for the Generation Time Data of Neumeyer (1992) at the Temperatures Observed.**

Temperature (°C)	Observed Gen. Time range	Square root predictions			Response Surface predictions	Davey predictions
		Usual fit	GLR (inv. Gauss.)	GLR (gamma)		
12.5	349-978	675	888	786	1094	1193
15.0	320-440	327	389	365	530	550
20.0	89-287	127	138	136	159	158
27.0	68	54	56	57	51	51
28.0	37-58	49	51	52	46	46
30.0	36-48	41	42	43	38	38
35.0	24-40	28	28	29	30	30

this (see also Baskerville, 1972; and Sprugel, 1983) is included in Table 4.12. The inverse Gaussian distribution is the least well-known of the three non-normal distributions considered here, its properties having been elucidated more recently than those of the other two distributions (see Folks, 1983). The variance-stabilising transformation $1/\sqrt{\mu}$ also needs an accompanying correction for bias, and although it is more difficult to find a published reference to this, an approximate bias-corrected estimate of the response time, θ, derived from a Taylor-series expansion by S.G. Candy (*pers. comm.*), is presented in Table 4.12 also.

It is interesting to note the agreement between the variance-stabilising transformations in Table 4.12 and formulae relating the variance on the time scale and the variances on the $\sqrt{k}$ and ln k scales that were presented in Ratkowsky *et al.* (1991). Those formulae are as follows:

$$V(\theta) = 4\theta^3 \ V(\sqrt{k}) \qquad (4.11)$$

Table 4.11 **Prediction of Response Times by Eqns. 4.8 - 4.10 for the Generation and Lag Time Data of Chandler (1988) and Smith (1985) at the Lowest Temperatures Observed.**

Data Set	Temp. (°C)	Observed Response Time	Square Root predictions: Usual Fit	Square Root predictions: GLR (inv. Gauss.)	Square Root predictions: GLR (gamma)	Response Surface predictions	Davey predictions
C*: 1a	4.4	1746	1850	1856	1832	1755	1805
C: 1b	4.4	260	279	281	273	265	271
C: 1c	4.4	1200	1267	1276	1236	1197	1283
C: 2	9.6	377	368	369	375	373	375
C: 3	10.1	409	436	438	434	431	432
S†: gen	10.0	6.68	7.87	8.02	7.09	6.78	7.03
S: lag	10.0	23.25	16.9	17.48	21.44	20.21	20.65

* C denotes Chandler data set, time in minutes. † S denotes Smith data set, time in hours.

and

$$V(\theta) = \theta^2 \ V(\ln k) \qquad (4.12)$$

respectively. Eqn. 4.11 tells one that if the variance is homogeneous in $\sqrt{k}$, justifying the fitting of the square root model in the form of Eqn. 4.8, then the variance of a time response θ is proportional to the cube of θ, as in the inverse Gaussian distribution, the scale parameter c equalling the constant $4V(\sqrt{k})$. Eqn. 4.12 tells one that if the variance is homogeneous in ln k, justifying the fitting of the response surface and Davey models in the form given by Eqns. 4.9 and 4.10, respectively,

Table 4.12 **Summary of the Form of Distribution, the Variance-stabilising Transformation, and the Formula for Obtaining the Bias-corrected Estimate of the Mean of the Response Variate on the Time Scale**

Distribution	Form of distribution	Transformation	Back-transformed estimate of θ, corrected for bias
Normal	$V = c$	$\upsilon = \mu$	No back transformation needed
Poisson	$V \doteq c\mu$	$\upsilon = \sqrt{\mu}$	$\hat{\upsilon}^2 + \sigma_R^2$
Gamma	$V = c\mu^2$	$\upsilon = \ln \mu$	$\exp\left(\hat{\upsilon} + \frac{1}{2}\sigma_R^2\right)$
Inverse Gaussian	$V = c\mu^3$	$\upsilon = 1/\sqrt{\mu}$	$\hat{\upsilon}^{-2}[1 + 3\sigma_R^2/(\hat{\upsilon})^2]$

[Note: σ_R^2 refers to the residual mean square, which is estimated in each case as the sum of squares of deviations of the *transformed* observed data from the fitted regression model, divided by the degrees of freedom, the latter being the number of data points minus the number of parameters estimated].

then the variance of a time response θ is proportional to the square of θ, as in the gamma distribution, with the scale parameter c equalling the constant V(ln k). This equivalence is not surprising since the derivation of those formulae is based upon the same Taylor-series expansion that was used originally to derive the variance-stabilising transformations. A potential use of relationships such as Eqns. 4.11 - 4.12 is to predict the variance of θ at temperatures for which there is no replication from temperatures in which the experimental conditions are replicated.

4.4.3. The Deterministic and Stochastic Components of a Regression Model

The two preceding sections show that there are two approaches for fitting the models given by Eqns. 4.8 - 4.10. One of these, based upon the use of generalised linear regression modelling, has the advantage of working directly with time, the response variable of interest, but requires specialised software for fitting the models. The second approach, via variance-stabilising transformations, converts the response variable (time) into a new variable that is approximately normally distributed, and can be carried out using one of the many widely available statistical software packages.

Clearly a regression model consists of both a *deterministic* component which relates a response variable, in this case a time θ, to the explanatory variables such as temperature (and pH, water activity, etc.) and a *stochastic* component which expresses the variability of the response variable or a transformation of that variable. The existence of a convenient deterministic relationship between explanatory and response variables does not necessarily carry any implications regarding the nature of the associated stochastic component. Thus, the use of ln θ, as in Eqns. 4.9 and 4.10, may be justified in that a simple linear regression model results, but this does not mean that ln θ is normally distributed. For some sets of data, the assumption that ln θ is normally distributed will be a good one. This is the case for the data of Smith (1985) for lag time, the means at each temperature being given in Appendix 4A.2. From the original full set of data made available to us, courtesy of M.G. Smith, CSIRO Division of Food Processing, Meat Research Laboratory, Cannon Hill, Queensland, Australia, it was found that the variance of lag time increased as the square of the mean, suggesting either the use of a logarithmic transformation before fitting that data set using ordinary linear regression, or to fit θ by generalised linear regression assuming gamma error. For the generation time data of Smith (1985), also made available to us by M.G. Smith, or those of Neumeyer (1992) in Appendix

4A.4, the more appropriate assumption for the error distribution of θ is one in which the variance is close to being proportional to the cube of the mean rather than to the square of the mean. This enables two courses of action to be taken. One may use generalised linear regression with the appropriate error assumption (inverse Gaussian), or one may rearrange the model to make use of the appropriate variance-stabilising transformation, which is to use $\sqrt{1/\theta}$ which is, of course, precisely the form used by the 'usual' approach when a model such as Eqn. 4.8 is employed.

If one re-examines the results in Table 4.10, it can be seen that the 'usual fit' by the square root model to the data of Neumeyer is consistently lower than the GLR predictions assuming inverse Gaussian error, the appropriate error assumption for those data. The underestimation is due to the fact that no bias correction was made; incorporating the bias correction, the formula for which appears in Table 4.12, will result in close agreement between the two sets of predictors. We note further that use of GLR assuming gamma error, the 'wrong' error assumption for Neumeyer's data, still produces reasonable estimates, suggesting a certain robustness about the estimation process.

4.5. SUMMARY

If replicated data are available, one should attempt to detect which is the appropriate stochastic assumption, and to base one's modelling procedure upon that assumption. Suppose one finds, from replicated data, that the variance, V, is proportional to the cube of the mean μ. This suggests that an inverse Gaussian error may be appropriate, resulting in two courses of action, *viz.* (i) use of GLR with that error assumption, or (ii) transformation of the model so that the response variable is $\sqrt{1/\theta}$. Thus, if one wishes to model the process using Eqns. 4.8 - 4.10, one would use Eqn. 4.8 in precisely the form as presented in 4.4.1., that is, as:

$$\sqrt{1/\theta} = a + bT$$

but one would have to rearrange Eqns. 4.9 and 4.10 to look as follows:

$$\sqrt{1/\theta} = 1/[\exp(a + bT + cT^2)]^{1/2} \qquad (4.9')$$

and:

$$\sqrt{1/\theta} = 1/[\exp(a + b/T + c/T^2)]^{1/2} \qquad (4.10')$$

respectively. These latter forms have the inconvenience of being nonlinear expressions, but they can nevertheless be fitted by nonlinear regression techniques which are intended for problems having a normal error distribution.

In a similar vein, suppose one found from preliminary investigation of replicated data sets that the variance was proportional to the square of the mean, or, equivalently, that the standard deviation was proportional to the mean. That would suggest that a gamma error assumption could be made, giving two alternative courses of action, *viz.* (i) use of GLR with that error assumption, or (ii) transformation of the model so that the response variable is ln θ. Thus, if one chooses Eqns. 4.8 - 4.10 as the models, one would use Eqns. 4.9 and 4.10 in precisely the same forms presented in 4.4.1., that is, as:

$$\ln \theta = a + bT + cT^2$$

and

$$\ln \theta = a + b/T + c/T^2$$

respectively, but Eqn. 4.8 would have to be re-arranged and fitted in the following form:

$$\ln \theta = -2 \ln (a + bT) \qquad (4.8')$$

Thus, one may still speak of a model being the 'square root' model, the 'response surface' model or the 'Davey' model, even if the forms in which the models appear seem unfamiliar. The important thing is to deduce the correct stochastic assumption for the data at hand, and to model accordingly.

As research proceeds with increasing automation enabling highly replicated experiments to be undertaken, it will become apparent whether one error distribution is appropriate to all data sets. In the meantime it must not be assumed that the error term behaviour selected for one experiment will necessarily apply to any other.

APPENDICES 4.

4A.1. Observed Values and Fitted Values for Various Models [Eqns. 3.10, 3.14 and 4.1] for *Pseudomonas* spp. Strain E5.2. [Data of Chandler (1988)].

(a) Data Set 1a: Time for change in optical density of 0.3, observed and fitted times (in minutes).

Temperature (°C)	Observed reaction time	Fitted Square- root Eqn. (3.10)	Fitted Eyring Eqn. (3.14)	Fitted Johnson & Lewin Eqn. (4.1)
4.4	1746	1850	1318	1743
9.6	758	739	753	781
11.2	628	597	636	625
12.4	561	515	562	534
13.6	460	449	496	459
15.0	379	387	430	389
16.6	331	331	366	327
17.8	291	296	324	290
19.1	249	265	285	257
20.3	234	240	253	232
21.6	203	216	223	210
23.2	192	192	191	188
24.5	176	175	168	173
25.7	164	162	150	162
26.8	156	151	135	154
28.1	141	139	120	145

(b) Data Set 1b: Generation time, observed and fitted times (in minutes)

Temperature (°C)	Observed reaction time	Fitted Square- root Eqn. (3.10)	Fitted Eyring Eqn. (3.14)	Fitted Johnson & Lewin Eqn. (4.1)
4.4	260	279	208	258
9.6	124	122	123	127
11.2	103	100	105	104
12.4	94	87	94	90
13.6	83	77	84	79
15.0	60	67	73	68
16.6	55	58	63	58
17.8	58	52	50	46
20.3	40	42	45	41
21.6	35.8	35.1	40	38
23.2	33	34.4	34	34
24.5	30.8	31.5	31	31
25.7	29.4	29.1	27	29
26.8	28	27.2	25	28
28.1	26	25.2	22	26

(c) Data Set 1c: Lag time, observed and fitted times (in minutes)

Temperature (°C)	Observed reaction time	Fitted Square- root Eqn. (3.10)	Fitted Eyring Eqn. (3.14)	Fitted Johnson & Lewin Eqn. (4.1)
4.4	1200	1267	880	1203
9.6	497.6	481	493	514
11.2	411.7	385	414	406
12.4	363.6	330	364	343
13.6	285.7	287	320	293
15.0	253.0	246	276	246
16.6	215.5	209	233	206
17.8	169.2	187	206	182
19.1	146.1	166	180	160
20.3	150	150	159	145
21.6	127.8	135	139	131
23.2	122.7	120	119	117
24.5	111.3	109	104	108
25.7	102.3	100	93	101
26.8	97.2	93.4	83	96
28.1	86.4	86.0	73	90

(d) Data Set 2: Lag time, observed and fitted times (in minutes)

Temperature (°C)	Observed reaction time	Fitted Square- root Eqn. (3.10)	Fitted Eyring Eqn. (3.14)	Fitted Johnson & Lewin Eqn. (4.1)
9.6	377	374	337	383
10.9	326.5	308	294	312
12.2	252	258	257	258
13.1	218	231	234	228
14.9	190	187	194	183
15.9	170.5	168	176	163
17.6	131.6	142	148	138
18.3	128.3	133	138	129
19.5	113.7	119	122	116
20.9	101.5	105	107	104
22.1	98.5	96	95	95
23.1	91.5	89	86	88
24.4	79.5	80	76	81
25.4	75.2	75	69	76

(e) Data Set 3: Lag time, observed and fitted times (in minutes)

Temperature (°C)	Observed reaction time	Fitted Square- root Eqn. (3.10)	Fitted Eyring Eqn. (3.14)	Fitted Johnson & Lewin Eqn. (4.1)
10.1	409	436	401	419
11.3	383	372	358	368
12.7	296	314	313	317
14.0	292.8	271	277	277
15.3	245.5	237	246	242
16.8	218.8	205	214	208
17.8	181.5	187	195	188
18.9	167	169	177	170
20.1	146.5	153	158	152
22.7	123.5	125	126	123
23.9	116.3	115	113	113
25.0	105.2	106	103	106
26.0	101.1	100	94	102

4A.2. Observed Values and Fitted Values for Various Models for Coliforms Growing in Meat. [Data of Smith (1985)].

(a) Data Set 1: Generation time data (times in hours)

Temperature (°C)	Observed reaction time	Fitted Square- root Eqn. (3.10)	Fitted Eyring Eqn. (3.14)	Fitted Johnson & Lewin Eqn. (4.1)
10	6.68	7.87	5.08	6.41
15	2.63	2.61	2.86	2.90
20	1.49	1.29	1.64	1.42
25	0.80	0.76	0.96	0.77
30	0.47	0.51	0.57	0.48
35	0.34	0.36	0.34	0.34
40	0.28	0.27	0.21	0.28

(b) Data Set 2: Lag time data (times in hours)

Temperature (°C)	Observed reaction time	Fitted Square- root Eqn. (3.10)	Fitted Eyring Eqn. (3.14)	Fitted Johnson & Lewin Eqn. (4.1)
10	23.25	16.93	13.24	23.06
15	5.90	6.22	7.40	6.08
20	3.00	3.20	4.21	2.87
25	1.92	1.95	2.44	1.85
30	1.19	1.31	1.44	1.31
35	0.96	0.94	0.87	0.96
40	0.74	0.71	0.53	0.71

4A.3. Observed and Predicted Lag Time for the Growth of Gram-Negative Spoilage Biota in Air-Packed Beef Mince. [Data of Adair *et al.* (1989)].

Lag time in hours.

Temperature (°C)	Observed data	Square Root (Eqn. 3.10) prediction	Square Root (Eqn. 3.10) Pearson Residual	'Schoolfield' (Eqn.3.20) prediction	'Schoolfield' (Eqn.3.20) Pearson Residual
-2.0	165.91	222.58	-0.01717	158.35	0.00379
0.0	98.38	114.84	-0.01337	107.08	-0.00785
2.0	85.37	69.89	0.02649	73.93	0.01800
5.0	36.19	39.53	-0.01344	44.33	-0.02758
10.0	20.83	19.80	0.01169	21.50	-0.00672
13.0	20.74	14.33	0.11816	15.04	0.09772
17.0	8.62	9.96	-0.04263	10.12	-0.04659
20.0	7.37	7.87	-0.02265	7.88	-0.02306
25.0	5.54	5.62	-0.00600	5.54	0.00000
30.0	4.25	4.20	0.00581	4.11	0.01680
Pearson goodness-of-fit			0.01785		0.01374
Deviance goodness-of-fit			0.01387		0.01166

4A.4. Generation Time Data for the Growth of *Staphylococcus. aureus* 3b. [Data of Neumeyer (1992)].

Temperature (°C)	Generation Time	Temperature (°C)	Generation Time
12.5	978.4	30	48.1
12.5	349.0	30	36.5
12.5	588.7	30	44.5
		30	35.9
15	325.7	30	44.9
15	440.0		
15	320.4	35	39.9
15	330.7	35	32.7
		35	36.4
20	192.9	35	37.1
20	260.8	35	35.0
20	286.7	35	29.1
20	211.0	35	28.0
20	89.1	35	28.4
20	110.4	35	28.4
20	130.4	35	24.3
20	235.0	35	28.1
		35	27.0
27	68.1	35	26.7
		35	27.2
28	57.6	35	24.6
28	41.2	35	29.5
28	37.1		
28	39.1		
28	38.2		
28	37.2		
28	45.3		
28	44.2		
28	39.2		
28	42.8		
28	44.2		
28	43.5		
28	45.1		
28	41.7		
28	44.8		
28	39.9		
28	40.8		
28	40.9		

Whilst this Chapter has been concerned primarily with modelling the effect of temperature on bacterial growth rate, other organisms display the same general response to temperature with rates increasing from a minimum to an optimum temperature and decreasing to zero in the superoptimal region. For completeness, the Arrhenius and Bělehrádek approaches are compared below for the growth of phytoplankton (Li and Dickie, 1987) and pollen tube penetration (Jefferies and Brain, 1984).

4A.5. Comparing Bělehrádek-type and Arrhenius-type Models Using Data on Microbial Plankton from Li and Dickie (1987)

For modelling the temperature dependence of the bacterial growth rate constant (or lag phase duration or generation time), the available evidence, some of which is presented in the body of this chapter, seems to favour the Bělehrádek-type models over the Arrhenius-type models, as the former class of models fit the data very well with a minimum number of parameters. In this Appendix we examine four data sets of Li and Dickie (1987) showing the dependence of the photosynthetic and heterotrophic activity of microbial plankton on temperature. These are suitable data sets for model comparison, as they combine accuracy of reaction rate measurement with the fact that those authors obtained rate measurements even at temperatures where the rates of incorporation were low. Because the data sets deal with plankton rather than bacteria, they enable one to examine whether the models used to describe bacterial growth are also suitable for other organisms. There was a total of four data sets presented in Fig. 3 of Li and Dickie (1987), the data having their origin in water samples collected from Bedford Basin, Nova Scotia. These data sets, representing activity-temperature curves obtained as a result of an experiment they conducted on 24 July 1986, were kindly made available to us for study by Dr. W.K.W. Li. These four activity-temperature curves for uptake of carbon and hydrogen, *viz.* total ^{14}C, ^{14}C-protein, total ^{3}H (tritium), and ^{3}H-protein incorporation, were typical of the data sets obtained from other experiments conducted by those authors. All had a similar shape, being unimodal with inflection points on both sides of T_{opt} [see Fig. 4A5.1 and Li and Dickie, (1987), Figs. 3 and 4].

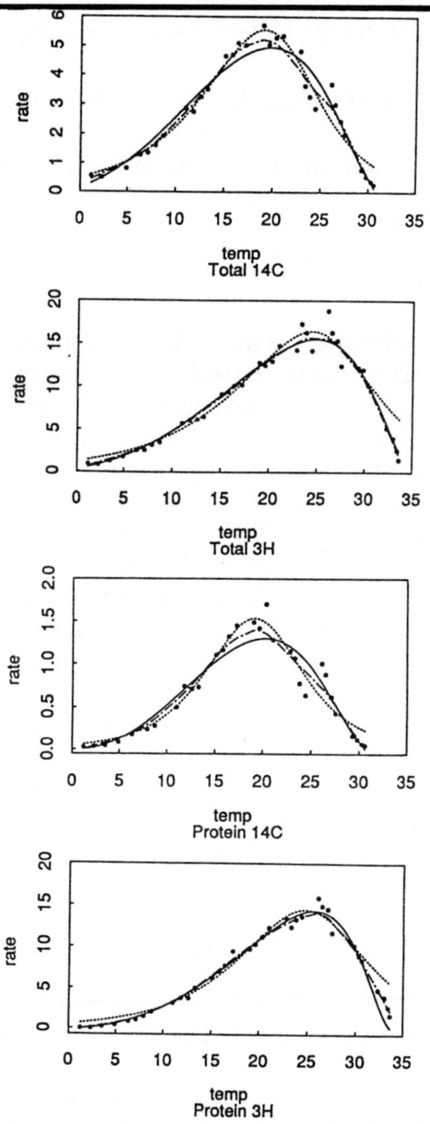

Fig. 4A.5.1 Comparison of three different models on the data of Li and Dickie (1987), 24 July 1986. Fitted values correspond to the four-parameter square root model, Eqn. 3.12 (solid line); the Johnson and Lewin model, Eqn. 3.16 (dotted line); and the "supsmu" smoothing function (dot-dash line). Temperatures shown are Celsius.

Li and Dickie (1987) studied a wide variety of mathematical models, which included the four-parameter square root model, Eqn. 3.12:

$$\sqrt{k} = b(T - T_{min})\{1 - \exp[c(T - T_{max})]\}$$

and the four parameter Johnson and Lewin model, Eqn. 3.16:

$$k = \frac{c \cdot T \cdot \exp(-\Delta H^{\ddagger}/RT)}{1 + \exp(\Delta S/R) \cdot \exp(-\Delta H/RT)}$$

They also fitted and tabulated results for four additional models, including a three-parameter curve based on the normal probability density function and a six-parameter Arrhenius-type curve incorporating an abrupt change of slope in the graph of ln *rate* versus inverse absolute temperature in the suboptimal temperature domain. None of these four additional curves fitted the data well, and were excluded from further consideration by Li and Dickie (1987) in their examination of the seasonal variation in temperature characteristics. Again, it must be remembered that a model consists both of a **deterministic** part [that is, the mathematical formula relating the explanatory (independent) and response (dependent) variables] and a **stochastic** part, which describes how the error term varies with changes in the explanatory variable (see Section 2.1.2.1). When the 'least squares' principle is used as the criterion of fit, as is almost universally done by users of regression models, different parameter estimates are obtained from different variants of the same model. Thus, Li and Dickie (1987), in employing the four-parameter square root model, actually fitted the form, in the present notation:

$$k = \{b(T - T_{min})\}^2\{1 - \exp[c(T - T_{max})]\}^2 \qquad (4A.5.1)$$

rather than the form given by Eqn. 3.12. The set of estimates $\hat{b}$, $\hat{T}_{min}$, $\hat{c}$, $\hat{T}_{max}$, obtained from least-squares fitting, will, in general, be different if Eqn. 4A.5.1 is used rather than Eqn. 3.12, and they may differ markedly. This results from a different stochastic assumption implicit in the use of these equations. Use of Eqn. 4A.5.1 in preference to Eqn. 3.12 for least-squares fitting would be justified if one believed that, as T varied, the errors in the response variable were homogeneous in k

rather than in $\sqrt{k}$, or, stated differently but equivalently, that the variance of k, namely V(k), rather than the variance of $\sqrt{k}$, namely V($\sqrt{k}$), was constant irrespective of temperature. Previous experience (Ratkowsky *et al.*, 1982, 1983) suggests that it is V($\sqrt{k}$) rather than V(k) that is homogeneous.

A similar problem occurs in the identification of an appropriate error term when fitting the Johnson and Lewin model. There is a choice, *inter alia*, between using the stochastic assumption implicit in writing Eqn. 3.16, where V(k) is assumed to be homogeneous, or in using the model after logarithmic transformation:

$$\ln k = \ln c + \ln T - \Delta H^{\ddagger}/RT - \ln[1 + \exp(\Delta S/R)\exp(-\Delta H/RT)] \qquad (4A.5.2)$$

in which the implicit assumption is that V(ln k) is constant as one varies temperature. As was the case with the use of Eqn. 4A.5.1 *versus* Eqn. 3.12, use of Eqn. 4A.5.2 in preference to Eqn. 3.16 may result in markedly different values of the parameter estimates $\hat{c}$, $\widehat{\Delta H}^{\ddagger}$, $\widehat{\Delta S}$ and $\widehat{\Delta H}$. In deciding between Eqns. 4A.5.1 and 3.12, and between Eqns. 3.16 and 4A.5.2, one would like to base the choice between models upon considerations such as goodness-of-fit and upon estimation behaviour (Bates and Watts, 1988; Ratkowsky, 1983, 1990), all the models in question being nonlinear regression relationships. Decisions based upon goodness-of-fit are not always easy to make, since, for example, in deciding between Eqns. 4A.5.1 and 3.12, one is in fact comparing the fit of k [for 4A.5.1] *versus* the fit of $\sqrt{k}$ [for 3.12], these being in effect two different responses. Similarly, comparing Eqns. 3.16 and 4A.5.2 involves decisions based upon the use of two incomparably different responses, *viz.* k and ln k, respectively. In nonlinear regression models, use of the correlation coefficient (or R^2 or adjusted R^2) as a measure of goodness-of-fit is not appropriate (Ratkowsky, 1983, 1990). One solution to the dilemma is to fit each pair of competing models using both stochastic assumptions, and to examine, after tabulation, the predicted values both as $\hat{k}$ and as $\widehat{\sqrt{k}}$ [for Eqns. 4A.5.1 *vs*. 3.12] or both as $\hat{k}$ and $\widehat{\ln k}$ [for Eqns. 3.16 *vs*. 4A.5.2].

The above procedure still has a basic problem; when the fit is based on assuming V(k) = constant, implicit in using Eqns. 3.16 and 4A.5.1, one will generally find that, because of the nature of the least-squares criterion, with its *modus operandi* of trying to minimise the sum of squares of the difference between observed and predicted values, the fitted values of $\hat{k}$ will tend to agree with the observed values of k better than the square roots of their $\hat{k}$ values will agree with the square roots of the observed k values. Similarly, the fit based upon assuming V($\sqrt{k}$) = constant, implicit in using Eqn. 3.12, results generally in fitted values $\widehat{\sqrt{k}}$ that agree with the observed values of $\sqrt{k}$ better than the $(\widehat{\sqrt{k}})^2$ values agree with the observed values of k. A similar phenomenon occurs when fitting the assumption V(ln k) = constant, implicit in using Eqn. 4A.5.2; the fitted values, $\widehat{\ln k}$ agree with the observed values of ln k better than they agree, after exponentiation, with the observed values of k. This tendency of the agreement appearing to be better when the comparison is made with $\hat{y}$ (say) with V(y) = constant as the stochastic assumption, in contrast to some other transformation of $\hat{y}$, say $\widehat{\sqrt{y}}$ or ln $\hat{y}$, is an inherent feature of the least-squares criterion. Thus, it may be more sensible to choose a 'neutral' basis of comparison, for example, by converting the fitted values of $\hat{k}$ [from Eqns. 3.16 or 4A.5.1] to $1/\hat{k}$, the latter being a measure of generation time or development time. Similarly, one can convert $\widehat{\sqrt{k}}$ [from Eqn. 3.12] to $1/(\widehat{\sqrt{k}})^2$ or $\widehat{\ln k}$ [from Eqn. 4A.5.2] to $1/\exp(\widehat{\ln k})$. The resulting converted values can then be compared with the observed values of $1/k$ and a decision made about which model gives the better fit. The fact that the data sets of Li and Dickie (1987) do not contain replicated temperatures prevented the approach described in 4.4 being attempted.

By use of the 'neutral' basis of comparison described in the above paragraph, it was determined that, for the four data sets of Li and Dickie (1987), Eqn. 3.12 was preferable to Eqn. 4A.5.1 for the square root model, and Eqn. 3.16 was preferable to Eqn. 4A.5.2 for the Johnson-Lewin model. Fig. A4.1 shows the observed values of k, the fitted values $\hat{k}$ from Eqn. 3.16, the transformed values $\hat{k}$ [using $(\widehat{\sqrt{k}})^2$] from Eqn. 3.12, and fitted values of $\hat{k}$ using the smoothing function "supsmu" in the statistical graphics computer package "S-PLUS". The "supsmu" function (Friedman, 1984) is an example of a smoothing algorithm that fits a smooth curve to a graph of one variable *versus* another variable with a minimum of distributional assumptions. It is included in Fig. 4A.1 to help readers make a decision about which of the competing models fits the data 'better'. For the total tritium and protein tritium data of Li and Dickie (1987), the four-parameter square root model, Eqn. 3.12, is a clear winner over the Johnson and Lewin model, Eqn. 3.16, the former model closely matching the data and also agreeing closely with the "supsmu" curve, except for protein tritium at temperatures above 30°C, where Eqn. 3.12 nevertheless provides a better fit to the data than Eqn. 3.16. For total carbon and protein carbon, the decision between models is less straightforward, except that Eqn. 3.12 is clearly preferable at temperatures above 25°C. In the middle temperature range of 15-25°C, which includes T_{opt}, the Johnson and Lewin model, Eqn. 3.16, has a slight edge over its rival, but "supsmu" mostly lies between the other two fitted curves, telling readers that the best-fitting model is somewhere 'in between' the two competitors.

When the comparison between models is made only on the low temperature region by fitting the low temperature data using the simple square root model, Eqn. 3.10:

$$\sqrt{k} = b(T - T_{min})$$

versus the Eyring (1935) model, Eqn. 3.14:

$$k = KT\exp(-\Delta H^{\ddagger}/RT),$$

Eqn. 3.10 emerges as a clear winner (see Fig. 4A.5.2) for all four data sets from Li and Dickie (1987), as that model fits the observed data better than Eqn. 3.14 and more closely matches the "supsmu" fit. A benefit of the form of Eqn. 3.10 is that the parameter T_{min} is a constant for the organism in question, irrespective of whether it is in reality a notional temperature or a realisable minimum temperature. In contrast, the Eyring model, Eqn. 3.14, not only does not fit data nearly as well as Eqn. 3.10, but has a parameter $\Delta H^{\ddagger}$ which is a function of temperature,

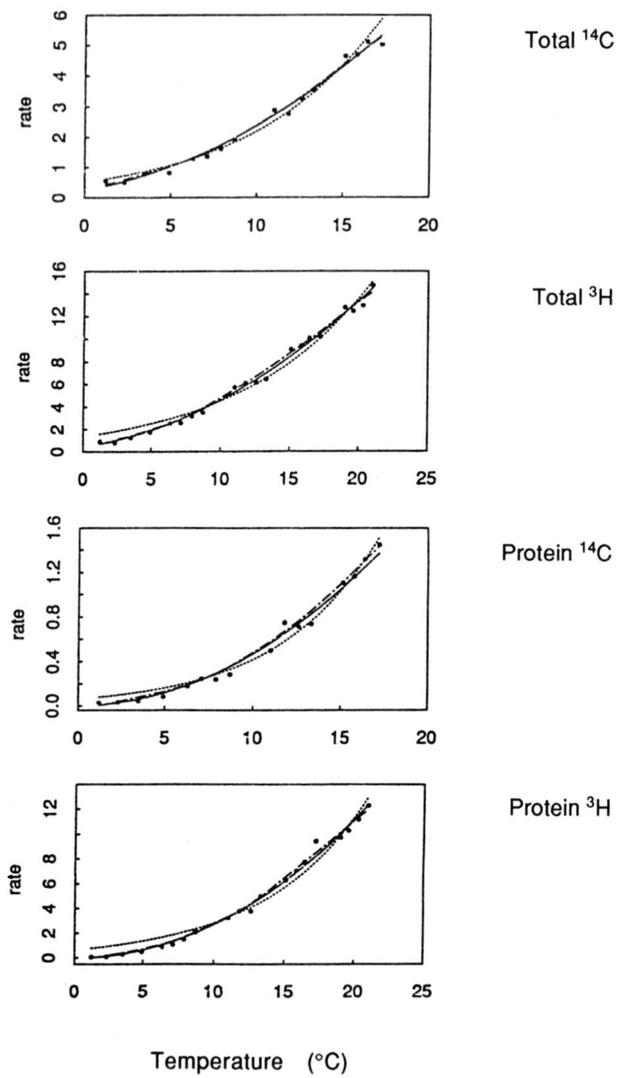

Fig. 4A.5.2 Comparison of three different models on the data of Li and Dickie (1987), 24 July 1986, suboptimal temperature data only. Fitted values correspond to the simple square root model, Eqn. 3.10 (——); the Eyring model, Eqn. 3.14 (·······); and the "supsmu" smoothing function (· — · — · — ·).

rendering it less useful than T_{min} for classifying or characterizing organisms.

For the overall temperature region the choice between the two major competing models, Eqns. 3.12 and 3.16, is less clear than in the low temperature region. The extension of the simple square root model, Eqn. 3.10, to give Eqn. 3.12, by inclusion of the factor:

$$1 - \exp[c(T - T_{max})]$$

was dictated by a need to find an expression to force √k to decline rapidly as temperature exceeds T_{opt} and approaches T_{max}. The advantage of a factor such as the above over a factor such as :

$$1 + \exp[(\Delta S - \Delta H/T)/R],$$

which appears in Arrhenius-type models, is that the former gives a zero rate at a finite temperature, $T = T_{max}$, whereas the latter expression declines to zero only when T approaches infinity (see Fig. 3.2). The latter failing is not merely a question of academic interest, as can be seen by examining the dotted curves in Fig. 4A.5.1, representing the fitted values of Eqn. 3.16. For all four data sets, the fitted values are remote from the data points and decline only slowly as temperature increases. A similar phenomenon occurs in the low temperature region, with $k \to 0$ as $T \to 0$, but the effect is less pronounced. Thus, Eqn. 3.16 gives unrealistic growth rate predictions at both low and high temperatures, a similar failing to that exhibited by the Sharpe and DeMichele (1977) model, as pointed out by Lowry and Ratkowsky (1983).

Not only does the four-parameter square root model, Eqn. 3.12, fit data well, it gives two cardinal temperatures T_{min} and T_{max} directly, with the third, T_{opt}, being easily derived once the model has been fitted. Table 4A.5.1. lists the three cardinal temperatures for the 24 July 1986 results of Li and Dickie (1987). Alternative estimates of T_{min} are provided by use of Eqn 3.10, which employs temperatures only in the low temperature region. These estimates are included in parentheses in Table 4A.5.1. The agreement between the two sets of estimates is reasonable though not perfect. The estimates obtained using Eqn. 3.10 are to be preferred, since they are based only on the data in the low temperature region.

Table 4A.5.1 **Estimates of Cardinal Temperatures for the 24 July 1986 Data of Li and Dickie (1987), Obtained Using the Four-parameter Square Root Model, Eqn. 3.12†**

	T_{min}	T_{opt}	T_{max}
Total Carbon	270.3 (268.6)	293.0	304.8
Protein Carbon	273.8 (273.1)	293.4	304.5
Total Tritium	270.2 (269.0)	298.1	308.5
Protein Tritium	273.7 (273.5)	298.9	307.1

† [T_{min} values in parentheses were obtained using Eqn. 3.10 in conjunction with sub-optimal temperature data only.]

4A.6. Fitting the Simple Square Root Model to the Data of Jefferies and Brain (1984) on Pollen Tube Penetration

Jefferies and Brain (1984) modelled isothermal growth curves for pollen-tube penetration within styles of flowers of the domestic apple (*Malus domestica* Borkh. cv. "Cox's Orange Pippin"). They found that, at each temperature within the range 3.5 - 33.5°C, pollen tube penetration could be modelled by a three-parameter Gompertz curve (Eqn. 2.9):

$$L(t) = C \exp\{ -\exp[-B(t - M)]\},$$

where:

C is the asymptotic pollen-tube length as t increases indefinitely
M is the value of t at which the absolute growth rate reaches its maximum value, and
B is the relative growth rate at that time.

A graph of pollen-tube penetration *versus* time, given in their Fig. 3 for 13.5°C, showed clear heterosedasticity in the data, the variance markedly increasing with increasing pollen-tube penetration. From their Materials and Methods section, there was no indication that this heterogeneity was taken into account in the estimation of the three parameters by, for example, use of a weighted regression. Values of estimates of C, B, and M at each of 12 temperatures studied by Jefferies and Brain (1984) are given in Table 4A.6.1.

A graph of the square root of B *versus* T (Fig. 4A.6.1) indicates that the simple (two parameter) square root model, Eqn. 3.10:

$$\sqrt{k} = b(T - T_{min})$$

could provide a good representation of the manner in which B changes with temperature. Hence, we fitted this model to the estimates of B in Table 4A.6.1 using linear least-squares to estimate a and b in:

$$\sqrt{k} = a + bT$$

and then determined T_{min} as $-a/b$. The resulting fitted model using the data for temperature and maximum relative growth rate B in Table 4A.6.1 is:

$$\sqrt{B} = 0.0764(T - 271.1)$$

Table 4A.6.1 **Temperatures and Estimates of Gompertz Function Parameters. Data of Jefferies and Brain (1984).**

Temperature (°C)	C (mm)	B (day^{-1})	M (days)
3.5	9.19	0.25	4.33
6.0	13.66	0.38	2.50
8.5	13.73	0.60	1.95
11.0	12.86	0.88	1.19
13.5	14.02	1.37	0.99
16.0	10.49	1.98	0.70
18.5	13.49	2.32	0.63
21.0	13.06	2.98	0.50
23.5	11.05	4.00	0.40
26.0	9.89	4.46	0.29
31.0	5.38	7.91	0.22
33.5	3.78	6.27	0.17

[*from* Jefferies and Brain (1984) by permission of Springer-Verlag, Berlin].

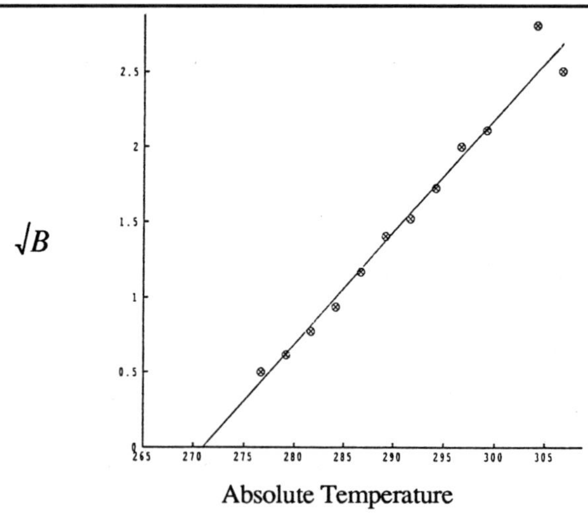

Fig 4A.6.1 Graph of $\sqrt{B}$ versus Temperature. [Data of Jefferies and Brain (1984)].

with an adjusted R^2 (percent variance accounted for) of 97.4%.

The graph of $\sqrt{B}$ *versus* T (Fig. 4A.6.1) shows considerable variability of the response at the two highest temperatures. This may be attributed to the failure of those authors to use a weighted regression, or it may be due to other factors, but there does not appear to be any systematic departure of the data from the simple square root model.

Since M is a time, one can take its reciprocal and also attempt to obtain a square root relationship with temperature. A graph of $\sqrt{1/M}$ *versus* T (Fig. 4A.6.2) indicates that a square root model should fit well. The fitted model is:

$$\sqrt{1/M} = 0.0624\ (T - 270.0)$$

with an adjusted R^2 (percent variance accounted for) of 98.5%, which is slightly higher than that obtained when modelling the square root of the maximum relative growth rate B. The T_{min} values are fairly close.

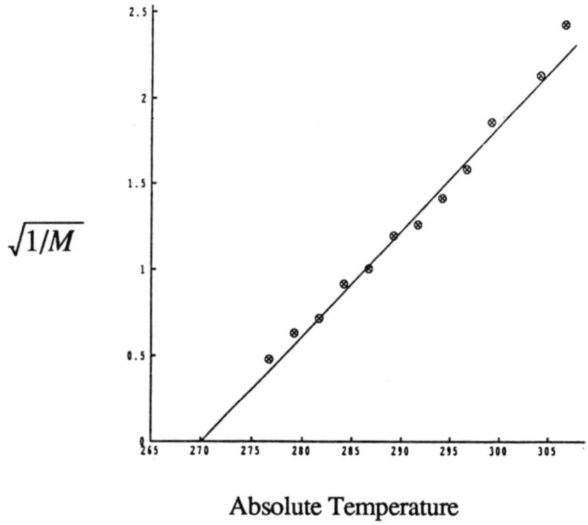

Fig 4A.6.2 Graph of $\sqrt{1/M}$ *versus* Temperature. [Data of Jefferies and Brain (1984)].

The graph of $\sqrt{1/M}$ *versus* T given in Fig. 4A.6.2 shows that the square root model adequately expresses the variation of M with temperature.

We note here that Jefferies and Brain (1984) expressed each of the three parameters, C, B and M of the Gompertz model as rather complicated nonlinear functions of temperature, to obtain a ten-parameter empirical model relating pollen penetration to temperature and incubation time. The approach we use here, in contrast, is based upon the use of simple, accurate mathematical models containing few parameters to make future predictions. These results show that the square root model is potentially useful for biological modelling even where bacteria are not the organisms of interest. Although the square root model does not apply to all modelling situations, a wide range of biological processes may have their temperature dependence modelled by a Bělehrádek-type model with exponents other than 0.5 on the rate parameter.

5.

MODELLING THE COMBINED EFFECT OF TEMPERATURE, WATER ACTIVITY AND OTHER FACTORS ON MICROBIAL GROWTH RATE

This chapter continues the comparison of models that was begun in Chapter 4, where temperature was the only changing factor considered. Here, in addition to temperature, the effect of water activity, pH, and nitrite and whether the atmosphere is aerobic or anaerobic is considered using real data sets. In addition to the Arrhenius-type and the Bělehrádek-type models considered in Chapter 4, the present chapter also considers polynomial relationships. The question of whether the individual factors are independent or interactive is explored.

5.1. INTRODUCTION

In Chapters 3 and 4, we considered situations in which temperature was the sole, or by far the major, factor controlling microbial growth rates. This situation pertains in foods such as fish, meat, milk and poultry which have high water activity levels and near neutral pH values. In many products, however, temperature is only one of a series of factors affecting the rate of product deterioration. The combined effect of different constraints was described as the 'Hurdle Concept' by Leistner and Rodel (1976), and updated by Leistner (1987).

The literature is replete with reports of the minimum values at which certain organisms (usually pathogens) will grow in laboratory media or food products. At intervals, the results of individual studies have been collated and compiled in review articles, for example, with respect to temperature, Elliot and Michener (1965), IUMS-ICFMH (1991) and for water activity, Scott (1957), Sperber (1983) and Beuchat (1987) provide comprehensive lists. The latter is concerned also with the water activities at which toxin production by pathogenic bacteria and fungi occurs.

The combined effect of different factors often receives attention, but again this information is usually provided in the form of discrete values of the hurdles under consideration that will not permit growth to occur. The general conclusion is that the range of one factor permitting growth will be greatest at the optimum level of a second factor.

While the type of information described above is clearly of considerable value, it has the disadvantage of being 'unique' and applicable only to the particular set of conditions, product, etc., under which it was obtained. In addition, Troller (1987) indicated that, while laboratory estimates of the minimum water activity ($\mathbf{a_w}$) for growth of *Staphylococcus aureus* are from 0.83-0.86, the 'casual' investigator would be hard pressed to obtain growth below $\mathbf{a_w}$ = 0.89 and in many foods below 0.90-0.91.

Troller (1987) argued further that "*it is more important to know the factors (including a_w) which impinge on microbial growth rather than placing too much confidence in a specific minimal number. Organisms in foods rarely encounter environments in which the a_w is poised at or below the minimal level for growth, and more frequently are faced with the need to grow and, even, prosper at levels intermediate between minimal and optimal*".

It appears that only on a few occasions have attempts been made to model the combined effects of two or more factors on the rate of growth, rather than state a particular combination of hurdles to prevent microbial development.

Davey (1989b) categorised modelling studies into those based on Arrhenius kinetics (Broughall *et al.*, 1983; Broughall and Brown, 1984); those based on the square root model (Chandler and McMeekin, 1989 a, b), and those employing polynomial models (Gibson *et al.*, 1988; Buchanan and Phillips, 1990). He indicated the latter are "*usually exercises in curve fitting, the results of which will remain unique, however immediately useful, since a difficulty is that the polynomial forms do not appear constant across a range of data, that is, they lack universality*". In this Chapter, we will compare these three approaches to modelling the combined effects of more than one constraint on microbial growth.

5.2. BĚLEHRÁDEK-TYPE RELATIONSHIPS

5.2.1. Temperature and Water Activity

A consistent set of relationships for describing the combined effect of temperature, *T*, and water activity, $\mathbf{a_w}$, the latter being altered by the addition of salt (NaCl), glycerol, sucrose or similar compounds, can be obtained using as a starting point the square root models of Ratkowsky *et al.* (1982, 1983), viz. Eqn. 3.10:

$$\sqrt{k} = b(T - T_{min})$$

and Eqn. 3.12:

$$\sqrt{k} = b(T - T_{min})\{1 - \exp[c(T - T_{max})]\}$$

Eqn. 3.10 was shown in Chapter 3 to be a special case of a general relationship advanced by Bělehrádek (1930), and has been found to be universally applicable for bacterial strains in describing the temperature dependence of the specific growth rate constant k in the low to suboptimal temperature region of growth, provided that growth conditions are non-limiting. The parameter T_{min} is characteristic of the organism (see Table 3.1 for specific examples) and the parameter b is a positive constant whose magnitude depends upon the units of k, the latter typically being hours^{-1} or minutes^{-1}. Eqn. 3.12 is an extension of Eqn. 3.10 to enable modelling of the temperature dependence of k to be extended to the whole of the biokinetic range. In using Eqns. 3.10 and 3.12, values of k, the specific growth rate constant, are obtained from cultures growing under conditions of unlimited nutrients. Thus temperature, T, forms the only limiting 'factor' to the growth of the bacterial system.

The organisms selected for detailed study by McMeekin *et al.* (1987) and Chandler and McMeekin (1989 a, b) were *Staphylococcus xylosus* and *Halobacterium* species isolated from salted, dried chub mackerel processed in Indonesia. The halotolerant coccus was capable of growth from 0 to 4 molal NaCl and 0 to 4.5 molal glycerol. The *Halobacterium* species required 3 molal NaCl to maintain the integrity of the cell and were able to grow in NaCl solutions up to 6 molal. For each of the strains tested in media adjusted by addition of NaCl or glycerol, the same response was observed. At each water activity, a square root plot provided a good description of experimental data, and extrapolation of individual regression lines to zero rate indicated a fixed T_{min} value (Fig. 5.1). The constancy of the theoretical T_{min} value was a bonus in the development of the combined temperature/water activity model. This meant that in the square root model only the constant b varied with water activity, that is, it was necessary simply to substitute b by a term describing the effect of water activity on rate.

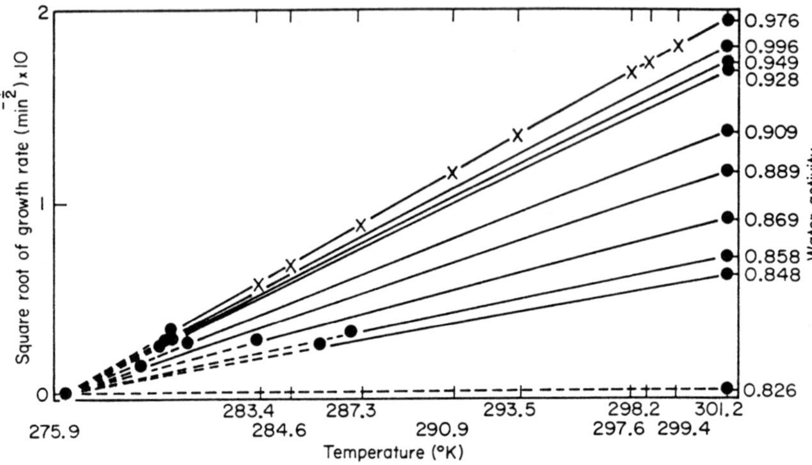

Fig. 5.1 Fitted lines for strain CM21/3 at nine different water activities at temperatures below T_{opt} using generation time data. Growth did not occur at $\mathbf{a_w}$ = 0.826 at temperatures below 28°C (301.2 K). The minimum temperature at which growth was measured at each $\mathbf{a_w}$ is indicated by •. [Adapted from McMeekin *et al.* (1987)].

For *S. xylosus* grown in NaCl or glycerol adjusted media, a plot of b^2 *versus* water activity was linear and from this the minimum water activity for growth in either humectant was estimated. From Eqn. 3.10, the basic square root equation, b^2 is proportional to rate, allowing substitution of b by a term for water activity to give an equation of the general form:

$$\sqrt{k} = d(T - T_{min})\sqrt{(\mathbf{a_w} - \mathbf{a_{w_{min}}})} \quad (5.1)$$

where $\mathbf{a_{w_{min}}}$ is a minimum water activity for growth.

Just as T_{min} may be a minimum temperature in the notional sense, the same may apply to $\mathbf{a}_{w_{min}}$. The parameter d is a constant for the particular organism and system in question. For *S. xylosus,* when common salt (NaCl) was used to adjust the values of water activity $\mathbf{a}_w$, the relationship:

$$\sqrt{k} = 0.159\ (T - 275.9)\ \sqrt{(\mathbf{a}_w - 0.838)} \qquad (5.2)$$

with k having units of h^{-1}, was found to apply to the water activity range 0.848-0.976 and to the temperature range between the minimum observed temperature at each $\mathbf{a}_w$ value to 28°C (301.2 K). The relationship does not apply for the value $\mathbf{a}_w = 0.996$ at which growth was slower than at $\mathbf{a}_w = 0.976$. For the same organism, when glycerol was used to adjust water activity, the relationship obtained was:

$$\sqrt{k} = 0.239\ (T - 276.6\ )\sqrt{(\mathbf{a}_w - 0.908)} \qquad (5.3)$$

with $\mathbf{a}_w$ ranging between 0.919 and 0.975 (Chandler and McMeekin, 1989a). The variation between the T_{min} values of 275.9 and 276.6 obtained from the two experiments is within the bounds of random variation and does not contradict the notion that T_{min} is an intrinsic property of the organism, in this case *S. xylosus*, but the clearly different values of 0.838 and 0.908 for $\mathbf{a}_{w_{min}}$ indicate that the minimum water activity is dependent upon the humectant employed.

With the halophilic *Halobacterium* species tested (Chandler and McMeekin, 1989b), the square root model was adequate to describe the effect of temperature and water activity constraints and again a fixed T_{min} value was implied. However, unlike the halotolerant *S. xylosus*, addition of NaCl beyond the minimum required for consistent growth of *Halobacterium* cultures had little effect on growth rate or the minimum temperature at which growth was observed. These are physiological features that might be inferred for bacteria totally adapted to life at elevated salt concentrations. Furthermore, they suggest that reasonable models to predict the growth of halobacteria at $\mathbf{a}_w$ levels below 0.85 could be based on the square root response to temperature without the necessity to incorporate a water activity parameter.

The growth rates of *S. xylosus* (McMeekin *et al.,* 1987), halobacteria (Chandler and McMeekin, 1989 a, b) and moulds (Wheeler *et al.*, 1986), have been incorporated into a spoilage envelope by Doe and Heruwati

(1988) which indicates graphically the $\mathbf{a_w}$ ranges in which growth of each is likely (Fig. 5.2, opposite).

Although Eqn. 5.1 applies only to the low temperature region, it seems reasonable to assume that an extension of Eqn. 3.12, incorporating a similar term to that in Eqn. 5.1 to account for the effect of water activity, *viz*:

$$\sqrt{k} = d(T - T_{min})\{1 - \exp\left[c(T - T_{max})\right]\}\sqrt{(\mathbf{a_w} - \mathbf{a_{w_{min}}})} \qquad (5.4)$$

may be suitable throughout the entire biokinetic temperature range. There are five parameters in this model: the two biological zeroes T_{min} and T_{max}, the minimum water activity $\mathbf{a_{w_{min}}}$, and the two system- and organism-dependent constants c and d. Later in this chapter we will investigate how well Eqns. 5.1 and 5.4 perform on additional sets of bacterial growth data that include the effect of changing $\mathbf{a_w}$ and T.

Temperature and water activity also combine at sub-zero temperatures to restrict the growth of some microorganisms. In this region water activity decreases progressively (Table 5.1), being dependent only on the vapour pressure of ice at any sub-zero temperature (Troller and Christian, 1978). This effect is significant in reducing the rate of growth of non-osmotolerant organisms, e.g. pseudomonads, and reduces their competitiveness with osmotolerant groups such as yeasts (Barnes *et al.*, 1978).

Organisms that are psychrotrophic and halotolerant are unaffected by water activity changes caused by partial freezing and continue to grow at a rate determined solely by temperature which can be modelled by the simple square root model to the point at which growth ceases (McMeekin and Franzmann, 1988).

5.2.2. Temperature and pH

The effects of pH, acidulant and temperature on the growth rate of *Yersinia enterocolitica* were reported by Adams *et al.* (1991) who proposed a model of the same general form as Eqn. 5.1 in which the $\mathbf{a_w}$ term was replaced by a pH term, *viz*:

$$\sqrt{k} = C\sqrt{pH - pH_{min}}\ (T - T_{min}) \qquad (5.5)$$

At every combination of pH and acidulant the response was well described by the square root model. The mean T_{min} value was

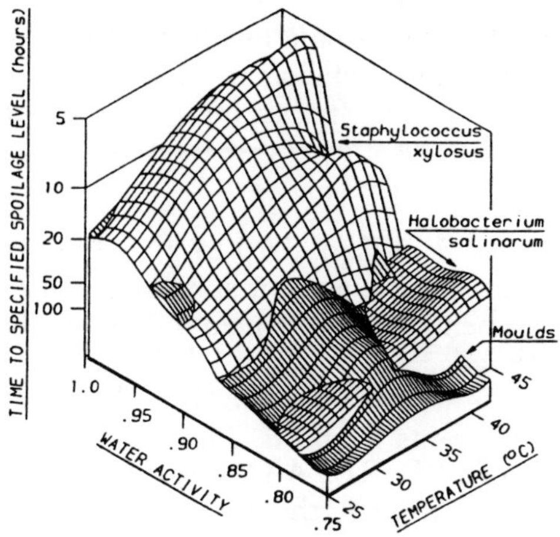

Fig. 5.2 Spoilage envelope - times to spoilage based on measured growth characteristics of certain moulds and bacteria. [Reproduced from Doe and Heruwati (1988) by permission of Applied Science Publishers, London].

Table 5.1 **Effect of Sub-Zero Temperature on Water Activity**

Temperature (°C)	Liquid Water (mmHg)	Ice (mmHg)	$a_w = \frac{\rho_{ice}}{\rho_{water}}$
0	4.579	4.579	1.00
- 5	3.163	3.013	0.953
- 10	2.149	1.950	0.907
- 15	1.436	1.241	0.864
- 20	0.943	0.776	0.823
- 25	0.607	0.476	0.784
- 30	0.383	0.286	0.75
- 40	0.142	0.097	0.68
-50	0.048	0.030	0.62

[*After* Troller and Christian (1978)].

269.0 ± 0.4 K and the constancy of T_{min} was unaffected by any of the combinations of acidulant and pH level examined (Table 5.2). Values of pH_{min} were obtained by plotting b^2 against pH; thus the pH-rate relationship is given by the expression:

$$b = C\sqrt{pH - pH_{min}} \tag{5.6}$$

which was substituted for *b* in Eqn. 3.10 to give Eqn. 5.5. The pH_{min} values obtained were acid-specific and for sulphuric, citric and lactic acid were identical to experimentally determined values of the lowest pH at which growth was observed.

5.3. ARRHENIUS-TYPE RELATIONSHIPS

Davey (1989b) used a modified Arrhenius-type model to describe the combined effect of temperature and water activity on bacterial growth rate. A precursor to the modified model, which was developed to describe irreversible thermal denaturation and which involved pH rather than water activity, was the expression:

$$\ln k = \Delta E/RT + a(\text{pH})^2 + b(\text{pH}) + c \tag{5.7}$$

where:

ΔE is an activation energy
R is the universal gas constant,
and a, b and c are additional parameters to be estimated.

This model was used successfully to describe combined pH and temperature effects on *Clostridium botulinum* inactivation and thiamine denaturation (Davey *et al.*, 1978). The modified form applied to temperature and water activity, which does not assume a straight-line relationship between ln *k* and reciprocal absolute temperature (1/*T*), is given by:

$$\ln k = C_0 + \frac{C_1}{T} + \frac{C_2}{T^2} + C_3\ \mathbf{a}_w + C_4\ \mathbf{a}_w^2 \tag{5.8}$$

Since the five parameters C_0, C_1, C_2, C_3 and C_4 all appear linearly, their estimation may be carried out using the widely used technique of

Table 5.2 ***T_{min}*** **values for** ***Yersinia enterocolitica*** **0:3 Estimated from the Square Root model, Eqn. 3.10**

Acid	pH	T_{min}	r
Acetic (pH_{min} = 4.8)	6.5	268.8	0.9974
	6.0	269.0	0.9951
	5.5	270.5	0.9913
Lactic (pH_{min} = 4.7)	6.5	268.0	0.9941
	6.0	268.0	0.9888
	5.5	268.4	0.9912
	5.0	267.7	0.9416
Citric (pH_{min} = 4.5)	6.5	268.6	0.9965
	6.0	268.8	0.9921
	5.5	269.0	0.9985
	5.0	269.0	0.9933
Sulphuric (pH_{min} = 4.0)	6.5	269.8	0.9952
	6.0	269.3	0.9965
	5.5	269.4	0.9989
	5.0	269.4	0.9954
	4.5	270.0	0.9894

[Adapted from Adams *et al.* (1991)].

multiple linear regression, available in most standard statistical computing packages. The absence of a 'cross-product' term involving $\mathbf{a_w}$ and T, such as $\mathbf{a_w}/T$, suggests that the effects of temperature and water activity on ln k are independent. The forms of Eqns. 5.1 and 5.4 also indicate independence of the effects of temperature and water activity terms on $\sqrt{k}$.

An important aspect of the form of Eqn. 5.8 is the inclusion of the term C_2/T^2 to model the curvilinear behaviour of the relationship between ln k and reciprocal absolute temperature $1/T$. Without that term, an expression incorporating additional terms to account for the effect of $\mathbf{a_w}$ would have the same failing as that of the simple Arrhenius model, namely, an inability to model adequately the temperature dependence when only temperature effects are involved.

The proposed model was evaluated using seven separate sets of published results from different investigators, including the data of Broughall *et al.* (1983) and McMeekin *et al.* (1987). Inclusion of results from these two groups is particularly useful as, together with the polynomial model of Gibson *et al.* (1988), it will permit direct comparison of the three major modelling approaches under consideration in this Chapter. Davey (1991) extended the application of his linear Arrhenius model to describe the lag phase of microbial growth in situations where temperature alone and temperature and water activity were the rate controlling factors.

Another approach, based upon Arrhenius-type kinetics to account both for temperature and water activity effects, is that of Broughall *et al.* (1983) who used a four parameter version of the Schoolfield *et al.* (1981) model, Eqn. 3.19:

$$1/g = \frac{\rho\,(25°\mathrm{C})\cdot\frac{T}{298}\cdot\exp\left[\frac{\Delta H_A^{\ddagger}}{R}\left(\frac{1}{298}-\frac{1}{T}\right)\right]}{1+\exp\left[\frac{\Delta H_L}{R}\left(\frac{1}{T_{1/2_L}}-\frac{1}{T}\right)\right]} \qquad (5.9)$$

where g = generation time, and all other symbols are as for Eqn. 3.19.

Readers should note that water activity never appears explicitly in the Schoolfield *et al.* (1981) model. Instead, the effect of changing water activity is to alter the magnitude of the four constants, ρ (25°C), $\Delta H_A^{\ddagger}$, ΔH_L and $T_{1/2_L}$, which, after some preliminary data-fitting attempts, were found by Broughall *et al.* (1983) to fit relationships such as:

$$\ln \rho\,(25°\mathrm{C}) = F_1 + G_1(a_W - 0.95),$$

$$\Delta H_A^{\ddagger} = F_1 + G_2(a_W - 0.95),$$

and so on, where the constants F_1, F_2, G_1, G_2, etc. are estimated by regression analysis. In Broughall and Brown (1984), the process is further extended to include the effect of changing pH. Those authors finished with a relationship having 14 parameters requiring estimation.

With both *S. aureus* and *Salmonella typhimurium* grown in UHT milk and glucose to adjust the water activity, Broughall and Brown (1984) found the Schoolfield model fitted generation time data better than lag time data. This was attributed to non-uniformity of the lag phase as a physiological process with significant microbial death occurring in some instances before exponential growth commenced.

Both test organisms behaved as expected with decreasing water activity with an increase in both generation time and lag time. The Gram-positive *S. aureus* was much less sensitive to lowered water activity than *S. typhimurium* which became non-culturable after a lag period of 5 days compared with 35 days for *S. aureus*. Also, the actual minimum temperature for growth occurred at higher temperatures with lowered water activity. This observation will be considered further in 5.6.

5.4. POLYNOMIAL RELATIONSHIPS

In this approach, temperature, water activity, pH and other physical and chemical properties of foods and their storage condition are modelled by use of a low-order polynomial. In a series of papers which included Gibson and Roberts (1986 a, b), Gibson *et al.* (1987, 1988), Roberts *et al.* (1981 a, b, c), that group of workers studied the effect of up to four combined factors on the growth of *Escherichia coli, Clostridium perfringens, Clostridium botulinum,* and a mixed inoculum of salmonellae. In each case, two of the factors were storage temperature and water activity (adjusted with NaCl) or pH, with the fourth factor, when used, being sodium nitrite.

Consider the situation such as pertained to the study reported by Gibson *et al.* (1988) which examined the growth responses of a mixed inoculum of *Salmonella thompson, S. stanley* and *S. infantis* at different temperatures, salt concentrations and pH levels. The database for predictions was obtained from determinations in laboratory media, and their derived model was then tested against the growth responses of salmonellae inoculated into a simulated cured pork product. The derived model was also compared with data from the literature. To eliminate subjective judgement of growth rates and lag time estimates the growth rate data were fitted to Eqn. 2.11, which in previous studies (Gibson *et al.,* 1987) was shown to give a better fit than Eqn. 2.10. Gibson *et al.*

(1988) found that as A, the lower asymptote, should be independent of conditions and that as no systematic variation of C, the difference between the upper and lower asymptote, could be observed with changing conditions, they were able to use mean values for these parameters in their model. This left B (the maximum exponential growth rate) and M (the time at which the growth rate is fastest) whose dependence on temperature (T), salt concentration (S) and pH (P) were modelled by polynomial expressions of the form:

$$\ln Y = a + b_1S + b_2T + b_3P + b_4S^2 + b_5T^2 + b_6P^2 + b_7ST + b_8SP + b_9TP + e \tag{5.10}$$

where:
Y is B or M, e is a random error,
and the coefficients $a, b_1, b_2, \ldots, b_9$ were determined in this multiple linear regression model by use of least squares.

The use of separate expressions for ln B and ln M implicitly assumes that these parameters are independent. Gibson *et al.* (1988) found that their approach provided a reasonable approximation to the underlying function.

Prior to their study of *Salmonella*, Gibson *et al.* (1987) had attempted to produce a similar model for growth of *Clostridium botulinum* in a pork slurry. However, because of difficulties encountered in accurate enumeration of the test organism, and the nature of the experimental system, a polynomial model was not presented. Instead, the data were shown as three-dimensional response surfaces. Response surfaces were also presented for *Salmonella* (Gibson *et al.*, 1988) and the effect of selective data removal on the resultant model and surface was discussed by Bratchell *et al.* (1989). Differences between response surfaces using reduced data sets illustrated differences between models and demonstrated the risk of producing an erroneous model by using insufficient data.

The polynomial/response surface approach has also been used by R.L. Buchanan and colleagues at USDA/ARS, Eastern Regional Research Center, Philadelphia. The outcome of a five-year program was to produce models describing the effect of five variables (temperature, NaCl level, pH, atmosphere, and $NaNO_2$ concentration) on the growth response of food-borne pathogens (*Salmonella*, *Shigella flexneri*, *Listeria monocytogenes*, *Staphylococcus aureus*, *Aeromonas hydrophila*, *Yersinia enterocolitica*, *Escherichia coli* 0157:H7 and *Bacillus cereus*).

A feature of the work of that laboratory is the production of a 'user-friendly' application program, the Pathogen Modelling Program (PMP), developed on the basis of a commercially available spreadsheet program, Lotus 1-2-3™ [1].

The general procedure employed is exemplified by Buchanan and Phillips (1990) who developed a model for *Listeria*. A factorial plus supplemental central composite experimental design (see 2.2.2.2.) was chosen to assess temperature effects from 5-37°C, pH values from 4.5-7.5, NaCl levels of 5-45 gL^{-1}, gaseous atmosphere and $NaNO_2$ levels of 0-1000 $\mu g\ mL^{-1}$.

A total of 709 growth curves were generated and fitted using Eqn. 2.11 and the results combined to generate a cubic model to predict *Listeria* growth in response to any combination within the range of variables listed above. The cubic model was used to take account of potential three-way interactions and fitted best when no-growth data were excluded. The complexity of the model generated can be readily appreciated from Table 5.3, overleaf, reproduced from Table 3 of Buchanan and Phillips (1990). Despite this complexity, the authors found that the model provided reasonable predictions of the behaviour of *Listeria monocytogenes* in food systems. As observed by other workers, predictions of lag phase duration were less accurate than generation time values, a fact attributed to the greater inherent variability of the physiological processes occurring during the lag phase and the inability to judge accurately the end of the lag phase.

Studies such as those of Buchanan and Phillips (1990) represent a monumental amount of work indicative of the type of time-consuming, painstaking effort required to generate sufficient data for predictive purposes. The USDA group is also to be commended for their open approach in offering their data bases for inspection, together with the PMP.

---

[1] **Lotus 1-2-3™**: Lotus Development Corporation, 55 Cambridge Parkway, Cambridge, MA 012142, USA.

Table 5.3 A Polynomial Model

Aerobic

$$
\begin{aligned}
\mathrm{Ln}(M) = {} & 37.657 + 0.0135T - 13.7331P + 0.4013S + 0.0713N + \\
& 0.00375T^2 + 1.9759P^2 - 0.000667S^2 - 0.000007051N^2 - \\
& 0.083TP + 0.000842TS - 0.0002141TN - 0.1155PS - \\
& 0.0167PN - 0.000125SN + 0.0000292T^3 - 0.0935P^3 + \\
& 0.00000328S^3 + 0.000286TPS + 0.0000315TPN + \\
& 0.00000014TSN + 0.0000175PSN - 0.000384T^2P - \\
& 0.00000855T^2S - 0.00000043T^2N + 0.00731TP^2 - \\
& 0.0000441TS^2 + 0.00672P^2S + 0.000968P^2N + \\
& 0.000294PS^2 + 0.00000062PN^2 - 0.00000016S^2N
\end{aligned}
$$

Degrees of freedom = 308
$R^2 = 0.967$

$$
\begin{aligned}
\mathrm{Ln}(B) = {} & -47.709 + 0.1631T + 18.6861P - 0.3609S + 0.01N - \\
& 0.00161T^2 - 2.7074P^2 + 0.00623S^2 - 0.0000863N^2 + \\
& 0.0242TPS - 0.000906TS + 0.000594TN + 0.0671PS - \\
& 0.00715PN + 0.000337SN - 0.0000648T^3 + 0.1276P^3 - \\
& 0.000029S^3 - 0.000551TPS - 0.0000733TPN - \\
& 0.00000033TSN - 0.0000431PSN + 0.000189T^2P + \\
& 0.0000549T^2S - 0.00000047T^2N - 0.00222TP^2 + \\
& 0.0000459TS^2 - 0.00000002TN^2 - 0.000781P^2S + \\
& 0.000777P^2N - 0.000872PS^2 + 0.0000112PN^2 - \\
& 0.00000038S^2N
\end{aligned}
$$

Degrees of freedom = 308
$R^2 = 0.942$

Anaerobic

$$
\begin{aligned}
\mathrm{Ln}(M) = {} & 89.9195 - 0.5378T - 38.8065P + 1.735S + 0.2175N + \\
& 0.00284T^2 + 5.9583P^2 + 0.00962S^2 - 0.000186N^2 + \\
& 0.1063TP - 0.00159TS + 0.000397TN - 0.567PS - \\
& 0.0574PN + 0.0000813SN - 0.0000321T^3 - 0.3024P^3 - \\
& 0.000107S^3 - 0.0000148TPS - 0.0000468TPN - \\
& 0.00000118TSN - 0.0000143PSN + 0.000397T^2P - \\
& 0.0000126T^2S - 0.00000184T^2N - 0.00964TP^2 + \\
& 0.0000487TS^2 + 0.0000001TN^2 + 0.0436P^2S + \\
& 0.0038P^2N - 0.000461PS^2 + 0.0000247PN^2 + \\
& 0.00000123S^2N - 0.00000001SN^2
\end{aligned}
$$

Degrees of freedom = 211
$R^2 = 0.974$

$$
\begin{aligned}
\mathrm{Ln}(B) = {} & -78.2567 + 0.7928T + 34.3598P - 0.913S - 0.4437N + \\
& 0.00218T^2 - 5.3119P^2 - 0.00394S^2 + 0.000233N^2 - \\
& 0.2134TP - 0.00174TS - 0.00094TN + 0.3002PS + \\
& 0.1272PNS - 0.00015SN + 0.0000274T^3 + 0.2693P^3 + \\
& 0.0000493S^3 + 0.000442TPS + 0.0000985TPN - \\
& 0.00000047TSN + 0.0000304PSN - 0.00104T^2P + \\
& 0.00000175T^2S + 0.00000584T^2N + 0.0194TP^2 - \\
& 0.0000318TS^2 - 0.00000011TN^2 + 0.0238P^2S - \\
& 0.00911P^2N + 0.000215PS^2 - 0.0000298PN^2 - \\
& 0.00000068S^2N - 0.00000003SN^2
\end{aligned}
$$

Degrees of freedom = 211
$R^2 = 0.944$

Cubic models for the effects and interactions of temperature, T (°C); initial pH, P; sodium chloride content, S (g/l); and sodium nitrite concentration, N (μg/ml), on the aerobic and anaerobic growth of *Listeria monocytogenes* Scott A, using Ln(*M*) and Ln(*B*) transformations in conjunction with variable combinations producing no growth responses being treated as missing values. [Reproduced from Buchanan and Phillips (1990) with permission of International Association of Milk, Food and Environmental Sanitarians, Inc.].

5.5. TESTING THE THREE APPROACHES FOR MODELLING COMBINED EFFECTS OF MORE THAN ONE FACTOR ON MICROBIAL GROWTH RATES

To examine the efficacy of Eqn. 5.8 for relating the specific growth rate constant to the combined effects of temperature and water activity, Davey (1989b) collected seven data sets from the literature. To provide consistent units for comparative purposes, Davey converted all of the tabulated or graphical data to have units of reciprocal hours (h^{-1}).

Results obtained by Davey (1989b) from his fitting of Eqn. 5.8 to the seven data sets are given in Table 5.4, which, in addition to showing the least-squares estimates of the parameters, identifies the microorganism, the source of the data, the ranges of temperature and water activities, the sample sizes, and presents the 'percentage variance accounted for'. This latter quantity, also known as 'adjusted R^2' in statistics, is widely interpreted as representing the proportion of the total variance (that is, mean square) in the response variable which is explained by the multiple linear regression model. That is, the percentage variance accounted for is defined as:

% variance accounted for = 100 (1- Error Mean Square/Total Mean Square)

where the Total Mean Square is the total sum of squares divided by its degrees of freedom, n - 1, with n being the sample size.

The magnitudes of the % variance accounted for in Table 5.4 indicate that the Davey model, Eqn. 5.8, fits the data well. An encouraging feature is the fact that each of the coefficients $C_0, C_1, ..., C_4$ has a consistent sign throughout the whole of the seven data sets.

The negative sign for C_2 reflects the fact that the simple Arrhenius model for temperature alone, *viz.* [see Chapter 3, Fig. 3.1 and Eqn. 3.13]:

$$\ln k = \ln A - E/RT$$

is not a straight line, but in reality is concave towards the $1/T$ axis. For the microorganisms of Table 5.4, the term C_2/T^2 is, in absolute magnitude, approximately one-half of that of the term C_1/T. If the simple Arrhenius model were a valid one, C_2 would be zero and the term would not need to appear in the model.

Table 5.4 Fit of Eqn. 5.8, Davey's 'Modified Arrhenius' Model, for Rate Coefficient During Microorganism Growth Phase

								Range		
Micro-organism	C_0	$C_1 \times 10^{-5}$	$C_2 \times 10^{-7}$	C_3	C_4	% Variance accounted for	*n*	T (°C)	a_w	Reference
Pseudomonas† on ox muscle	−919·54	1·6033	−2·3784	1317·7	−669	99·0	15	−1–30	0·960–0·993	Scott (1937)
Aerobacter aerogenes	−1595·4	3·1597	−4.8378	2149·2	−1069	94·0	18	15–37	0·950–0·975	Wodzinski & Frazier (1961, by Casolari *et al.* (1979))
Pediococcus cerevisiae strain PC4 on GTY media	−1406·0	2.9553	−4·5427	1862·2	−937	92.9	16	24–42	0·944–0·998	Lanigan (1963)
Clostridium botulinum type E strain 103 in NYG media	−2227·8	0·47991	−0·84488	4339·1	−2177	97·9	15	15–30	0·970–0·995	Ohye *et al.* (1967)
Microbacterium thermosphactum 22 on APT media	−1347·5	1.3712	−2·0789	2278·7	−1158	98·0	18	0–25	0·930–0·990	Brownlie (1969)
Salmonella typhimurium in UHT milk	−1904·3	2·8434	−4·3618	2935·7	−1495	96·9	28	9–30	0·94–0·98	Broughall *et al.* (1983)
Staphylococcus xylosus	−550·95	2·5162	−3·8381	285·69	−147	97·2	185	7–27	0·848–0·996	McMeekin *et al.* (1987)

* k in h^{-1}. *T* in degree absolute.
† Scott (1937) and Empey & Scott (1939) reported this organism as *Achrombacter* 7 but it was subsequently identified as *Pseudomonas* by Brown & Weidemann (1958).

[Reproduced from Davey (1989a) with permisssion of Blackwell Scientific Publications].

Inspection of Table 5.5 shows that the ratio $C_3 : C_4$ is virtually constant with the value -1.977 ± 0.02 (S.D.) strongly suggesting that the Davey model is overparameterised. The ratio $C_2 : C_1$ ranges from -0.0148 for the *Pseudomonas* data set to -0.0176 for the *Clostridium botulinum* data; its magnitude appears to be related to the temperature span of the data set. For the *C. botulinum* data, the parameter C_2, which accounts for the deviation of the response from Arrhenius model behaviour, is very small. This is to be expected as the data range is narrow, being from 13 to 28°C above T_{min} (see 3.5, Fig. 3.3).

Ross (unpublished) using a data set for the growth of *Aeromonas hydrophila* 3459, found that he could simulate the small variation in the ratio of $C_2 : C_1$ by sequentially removing data points from the extremes of the temperature range of the data set, and again fitting a Davey model (Eqn. 3.22) to the reduced data set (McMeekin *et al.*, 1992). The apparent constancy of the ratio $C_2 : C_1$ is further confirmation of the significance of a Bělehrádek power function throughout the biokinetic temperature range. The magnitude of the constants of the Arrhenius plot depend only upon the distance along the temperature axis (away from T_{min}) and not the value of temperature.

Results using the extended square root models, Eqns. 5.1 and 5.4, are given in Tables 5.6 and 5.7. The coefficients in Table 5.6 are for the model given by Eqn. 5.1, where the temperatures were confined to the low to suboptimal range. This parsimonious model, containing only three parameters, has '% variance accounted for' values which are very comparable to those obtained by Davey (1989b) using Eqn. 5.8. The T_{min} values, *viz.* 275.4 for *Clostridium botulinum* type *E*, 264.6 for *Microbacterium thermosphactum,* 278.0 for *S. typhimurium* and 275.9 for *S. xylosus* are all very reasonable values for these microorganisms. Fig. 5.1 presents the data for *S. xylosus* in graphical form, having been reproduced from McMeekin *et al.* (1987). The series of straight lines, one for each water activity, converges to the T_{min} value of 275.9 and emphasises the simplicity of Eqn. 5.1 for modelling the combined effect of temperature and water activity.

The coefficients in Table 5.7 are for the model given by Eqn. 5.4, where the highest temperatures used either exceed T_{opt} or, in the case of the data set of Wodzinski and Frazier (1961), are very close to T_{opt}.

The '% variance accounted' for values are comparable to those obtained by Davey (1989b) using Eqn. 5.8. The T_{min} and T_{max} values

Table 5.5 **Analysis of Fitted Davey Models**

Microorganism	$\lvert C_2/C_1 \rvert$	$\lvert C_3/C_4 \rvert$	Temperature Range (°C)	Temperature Range>T_{min}	T_{min} (K)
Pseudomonas	0.0148	1.97	31	10 - 41	262
Microbacterium thermosphactum	0.0152	1.97	25	9 - 34	264
Aerobacter aerogenes	0.0153	2.00	22	5 - 27	283
Staphylococcus xylosus	0.0153	1.94	21	5 - 25	275
Salmonella typhimurium	0.0153	1.96	20	4 - 25	278
Pediococcus cerevisiae	0.0154	1.99	18	26 - 44	271
Clostridium botulinum	0.0176	1.99	15	13 - 28	275

for *Pseudomonas* spp. and *Pediococcus cerevisiae* are reasonable estimates of these cardinal temperatures for these organisms, as is the T_{min} value for *Aerobacter aerogenes*. The T_{max} value for the latter (330.1 K) is clearly too high and may be attributed to lack of data at superoptimal temperatures.

With one exception, the minimum water activity values predicted by Eqns. 5.1 and 5.4 and shown in Tables 5.6 and 5.7 are expected for the organisms concerned, for example, the predicted value for *C. botulinum* type E was 0.963 and that observed by Ohye and Christian (1967) was 0.97. For *Salmonella typhimurium* growing in UHT milk adjusted with glucose, Broughall *et al.* (1983) found a minimum $\mathbf{a_w}$ value of 0.94, compared with a predicted value from Eqn. 5.1 of 0.937. The exception was the predicted value for *Pseudomonas* species (0.856), where no

Table 5.6 **Parameter Estimates for Eqn. 5.1 for Published Data**

Organism	d	$a_{w_{min}}$	T_{min}	% variance accounted for	Reference
Microbacterium thermosphactum	0.114	0.929	264.6	96.4	Brownlie (1969)
Staphylococcus xylosus	0.159	0.838	275.9	95.8	McMeekin *et al.* (1987)
Salmonella typhimurium	0.221	0.937	278.0	96.6	Broughall *et al.* (1983)
Clostridium botulinum	0.246	0.963	275.4	98.0	Ohye *et al.*(1967)

Table 5.7 **Parameter Estimates for Eqn. 5.4 for Published Data**

Organism	d	$a_{w_{min}}$	T_{min}	c	T_{max}	% variance accounted for	Reference
Pseudomonas spp.	0.067	0.856	262.0	0.186	308.6	99.3	Scott (1937)
Aerobacter aerogenes	0.686	0.948	283.0	0.019	330.1	91.7	Wodzinski and Frazier (1961)
Pediococcus cerevisiae	0.121	0.939	271.0	0.386	318.9	97.3	Lanigan (1963)

growth would be expected at an $\mathbf{a_w}$ value of <0.96. The Scott (1937) data set, however, was complete only for $\mathbf{a_w}$ = 0.993, had only three data at $\mathbf{a_w}$ = 0.98 and 0.97, and only one datum at $\mathbf{a_w}$ = 0.96.

Another source of data on the combined effect of more than one factor on the rate of microbial growth is that of Gibson *et al.* (1988) for a mixture of three species of salmonellae grown in tryptone soya broth (TSB) at different temperatures, salt concentrations and pH values. As described in the previous section, Eqn. 2.11 was used to model the $\log_{10}$ bacterial counts, with two of the parameters of the model being set equal to mean values, and the other two being fitted as second order polynomial functions of the three factors determining growth (Eqn. 5.12). Data presented in Table I of Gibson *et al.* (1988) provide 66 values of growth rate derived from using Eqns. 2.11 and 5.12 in the way described in 5.4. The temperatures used were 10, 15, 20, 25 or 30°C except for two values at 27°C. The salt concentrations were conveniently grouped into five categories 0-1, 1-2, 2-3, 3-4 and 4-5 (% w/v) and the pH values also could be grouped around the 'target' levels used by Gibson *et al.* (1988) of 5.6, 5.9, 6.2, 6.4 and 6.8. The calculated values of growth rate in each of the 125 cell combinations thus formed (many cells were empty), showed that growth rate exhibited no consistent trend with changing pH, suggesting that pH can be ignored as a factor, at least within the range 5.6 - 6.8, which is far removed from pH_{min}. Averaging over all pH values for each of the 25 cells formed from combinations of temperature and salt concentration, gave the summary shown in Table 5.8 (four cells are empty). The derived growth rates behave in the expected way by increasing with increasing temperatures. With increasing salt concentration, or, equivalently, decreasing water activity, the rates generally decrease as expected, but there are a few anomalies, especially for the two highest temperatures at the 0-1 NaCl concentration. Remembering that these growth rates are derived, rather than observed, the anomalies may be related to the fact that the parameters of Eqn. 5.10, although correlated, were implicitly assumed to be independent of each other by Gibson *et al.* (1988) when they employed separate polynomial expressions to evaluate the derived growth rate. Nevertheless, the above exercise leads to the conclusion that for this data set pH need not be taken into consideration in modelling growth rate, and that the growth rate appears to behave in a manner which is consistent with Eqn. 5.1, the modified form of the square root model, Eqn. 3.10.

Table 5.8 **Average Derived Growth Rates (h^{-1}) of Salmonellae as a Function of Temperature and Salt (NaCl) Concentration Using the Polynomial Modelling Procedure of Gibson *et al.* (1988)**

Temperature (°C)	NaCl Concentration (%w/v)				
	0 - 1	1 - 2	2 - 3	3 - 4	4 - 5
10	0.03	0.03	0.025	0.020	0.025
15	0.11	0.11	0.100	0.053	0.065
20	0.30	0.29	0.250	-	0.160
25	0.41	0.50	-	-	0.270
30	0.42	0.86	0.550	-	0.330

A test of whether Eqn. 5.1 fits data for the combined effect of different temperatures and water activities, can be made using the data tabulated in Tables IV and V of Gibson *et al.* (1988), which recorded *observed* generation times and times for a thousand-fold increase in salmonellae numbers in TSB, rather than *derived* values. These values are transcribed in Table 5.9.

Water activities were calculated from the observed % NaCl values using the expression from Robinson and Stokes (1949):

$$\log_{10}a_w = -0.007824 \times 2 \times \left(\frac{\%\ \text{NaCl}}{58.4}\right) \times 0.986 \qquad (5.11)$$

where the factor 2 represents the number of dissociated ions, 58.4 is the molecular weight of sodium chloride and 0.986 is the osmotic coefficient.

The $\mathbf{a_w}$ values so obtained were then multiplied by 0.996, the water activity of the basal medium, to give the equivalent $\mathbf{a_w}$ values shown in Table 5.9.

Fitting Eqn. 5.1 to the generation time data by nonlinear least squares regression resulted in the following expression, where g is generation time:

$$\sqrt{\frac{1}{g}} = 0.363\ (T - 279.4)\sqrt{(\mathbf{a_w} - 0.949)} \qquad (5.12)$$

The predicted values of the generation time from Eqn. 5.12 are tabulated in Table 5.9. Similarly, Eqn. 5.1 was fitted to the data on the time ϕ for a thousand-fold increase in salmonellae to give

$$\sqrt{\frac{1}{\phi}} = 0.0756\ (T - 277.9)\sqrt{(\mathbf{a_w} - 0.949)} \qquad (5.13)$$

The fitted values are tabulated in Table 5.9. Clearly, the fit of both Eqns. 5.12 and 5.13 to the data are very good and the estimate of T_{min} for this mixture, although varying by 1.5°C between 6.2°C (279.4 K) from Eqn. 5.12 to 4.7°C (277.9 K) from Eqn. 5.13, represents a realistic estimate for salmonellae. Both data sets result in the same predicted value of $\mathbf{a}_{\mathbf{w}_{min}}$, *viz.* 0.949.

The main conclusions to be drawn from the fitting of models for the combined effect of temperature and water activity to the actual data sets described in the preceding paragraphs, is that the Bělehrádek-type Eqns. 5.1 and 5.4 are very suitable for fitting such data. In practice, some users might prefer a model such as Eqn. 5.8, because it is a multiple linear regression model whose fitting can be carried out using any number of readily-available statistical computing packages. In contrast, models such as Eqns. 5.1 and 5.4, being nonlinear regression models, require solution by some iterative process. Many statistical computer packages, however, offer the capability for carrying out nonlinear regression analysis, one of which is the very widely available package SAS (1987). To assist potential users who might be interested in fitting Eqns. 5.1 and 5.4 to their data, sample SAS code is given in Appendix 5A.1 to this Chapter. The code includes expressions for the derivatives of the function with respect to the parameters, needed as part of the fitting process (see 2.4.).

Table 5.9 **Observed Generation Times and Times for a Thousand-Fold Increase in Salmonellae Numbers in TSB, and Predicted Values of Those Times Using Eqns. 5.12 And 5.13. Data of Gibson *et al.* (1988)**

Temp. (°C)	NaCl (%w/v)	a_w	Generation time (h) Observed	Generation time (h) Predicted (Eqn. 5.12)	Time for 1000x increase(h) Observed	Time for 1000x increase(h) Predicted (Eqn. 5.13)
10	0.82	0.991	8.9	12.8	176	151
10	1.32	0.988	17.7	13.8	237	162
10	2.66	0.980	16.2	17.3	249	204
10	3.70	0.974	17.8	21.5	288	253
10	4.56	0.969	18.4	26.8	394	316
15	0.82	0.991	2.7	2.4	37	40
15	1.33	0.988	2.6	2.5	41	43
15	2.64	0.980	3.2	3.2	47	54
15	3.75	0.974	6.6	4.0	85	66
15	4.56	0.969	4.2	4.9	98	83
20	0.77	0.991	1.0	1.0	14	18
20	1.32	0.988	1.0	1.0	19	19
20	2.67	0.980	1.2	1.3	20	24
20	4.50	0.969	1.9	2.0	36	38
25	0.82	0.991	0.6	0.5	8.5	10
25	1.32	0.988	0.6	0.6	10.7	11
25	4.06	0.972	0.9	0.9	16	19
25	4.50	0.969	1.0	1.1	20	21
30	1.32	0.988	0.3	0.3	9.6	7
30	4.50	0.969	0.8	0.7	14.2	14

Appendix 5A.1 includes two illustrative examples, one fitting Eqn. 5.1 to the generation time data set of Gibson *et al.* (1988) and one illustrating the fitting of Eqn. 5.4 to the data of Lanigan (1963). In both illustrations, examples of the outputs generated by SAS are given.

An equation such as Eqn. 5.1 represents an extension of the simple square root model Eqn. 3.10 to accommodate the effect of changing water activity. The use of the simple multiplying factor $\sqrt{(\mathbf{a_w} - \mathbf{a_{w_{min}}})}$ worked very well for all the organisms shown in Table 5.4 and also for *Staphylococcus xylosus* for $\mathbf{a_w}$ values up to and including 0.976 but not including $\mathbf{a_w} = 0.996$, whose b value in Eqn. (3.10) was lower than the corresponding b value for $\mathbf{a_w} = 0.976$. To accommodate an 'optimum' $\mathbf{a_w}$ value, on both sides of which growth is slower, analogous to the optimum growth rate with respect to temperature, one might consider a modification of the factor $\sqrt{(\mathbf{a_w} - \mathbf{a_{w_{min}}})}$ to one consisting of the form:

$$\sqrt{b_1^2 \; (\mathbf{a_w} - \mathbf{a_{w_{min}}})^2 + (\mathbf{a_w} - \mathbf{a_{w_{min}}})}$$

where one needs an additional term with an additional parameter, denoted here as b_1. Similar terms may be needed to describe the effect of other factors. Although pH was not demonstrated to be a factor of importance in the work reported by Gibson and Roberts and their co-workers on 'normal' food spoilage, terms involving pH would need to be included in models for use with acid treated food surfaces or silage production. For example, they could be appended to the basic square root model, Eqn. 3.10 in a similar manner to the way $\mathbf{a_w}$ appears in Eqn. 5.1, that is, as a multiplying factor (Adams *et al.*, 1991). For the case where the data cover the entire biokinetic temperature range, rather than being confined to the suboptimal temperature region, similar modifications may be made to Eqn. 3.12.

5.6. IMPLICATIONS FOR THE MODE OF INTERACTION OF DIFFERENT FACTORS

In the introduction to this Chapter, the Hurdle Concept (Leistner, 1987) was invoked to explain the interaction of different factors on the growth of microorganisms in foods. At least some of the hurdle combinations described by Leistner (1987) imply a synergistic effect where the combined effect of two factors acting together is greater than that of the sum of the two individual constraints.

With respect to temperature and water activity, however, Davey (1989b) suggested that the effect of these hurdles is independent rather than interactive. Chandler (1988) had also proposed this effect on the growth rate of *S. xylosus* as T_{min} was fixed in square root plots and only '*b*' (the slope of the regression line) varied with water activity. From Fig. 5.1 it is evident that the effect of sequential reduction in water activity is proportionally the same at any temperature, and as temperature decreases the magnitude of growth rate reduction due to lowered water activity is also decreased.

The effect may also be inferred from Eqn. 3.24:

$$\Delta E_{app} = \frac{2RT^2}{(T - T_{min})}$$

which indicates that ΔE_{app} (the apparent activation energy) depends mainly on the difference between the temperature, T, and T_{min}.

Where the effect of the two factors is interactive rather than independent is seen at the point where growth ceases, that is, although T_{min} (the theoretical minimum temperature for growth) is fixed, the *actual* temperature at which growth ceases increases with decreasing water activity. Presumably the two factors act independently on different physiological functions of the cell which combine eventually, for example, to cause a reduction in ATP levels or adenylate energy charge to the point where growth is impossible. This type of effect is illustrated in Fig. 5.1 for *S. xylosus*. The actual minimum temperature at which growth rates were measured at the optimum $\mathbf{a_w}$ (0.976) was 7.4°C whilst at $\mathbf{a_w}$ = 0.848 the minimum temperature was 12.6°C. Applying Eqn. 3.24, ΔE_{app}, in the absence of a water activity constraint, was 343 kJ mol^{-1}. At $\mathbf{a_w}$ = 0.848 the energy barrier to growth that the cells were able to overcome was reduced to 134 kJ mol^{-1}. This reduction in the ability of the cells to overcome the barrier imposed by temperature might be interpreted as approximately 200 kJ mol^{-1} of the energy available to the cell being diverted to the production of compatible solutes to counteract the effects of lowered water activity.

A similar response to the combined effect of temperature and water activity is suggested for the other organisms studied by Davey (1989b), *viz. Pseudomonas* spp., "*Aerobacter*" *aerogenes, Pediococcus cerevisiae, Clostridium botulinum* type E, "*Microbacterium*" *thermosphactum* and *Salmonella typhimurium*. Further evidence for a fixed T_{min} value at different NaCl concentrations comes from the work of Gibson *et al.* (1988) for growth of *Salmonella* serotypes in a laboratory medium (Table 5.8) and a similar response at different pH levels can be inferred from the data presented by Adams *et al.* (1991). That the value of T_{min} is apparently unaffected by pH and NaCl concentration is consistent with the contemporary paradigm of protein conformational dynamics which is presented in Chapter 10.

The apparent constancy of T_{min} raises the intriguing possibility that the combined effects of temperature (below T_{opt}), $\mathbf{a_w}$ (below $\mathbf{a_{w_{opt}}}$) and pH (below pH_{opt}) might be described by an equation of the following type:

$$\sqrt{k} = c\sqrt{(pH - pH_{min})}\sqrt{(\mathbf{a_w} - \mathbf{a_{w_{min}}})}(T - T_{min}) \qquad (5.14)$$

For a given humectant and acidulant it would be necessary only to determine the values of $\mathbf{a_{w_{min}}}$, pH_{min} and T_{min} in 3 separate experiments, and to fit the value of the parameter c, to derive a model for the combined effects of temperature, pH, and $\mathbf{a_w}$.

APPENDIX 5.

5A.1. A SAS Program for Fitting Data to Eqns. 5.1 and 5.4

The following three pages relate to a SAS program for fitting Eqn. 5.1 to the generation time data of Gibson *et al.* (1988). The first page gives the SAS code, the second and third pages give the output obtained using that code. Parameter estimates, their asymptotic standard errors, their correlations, and fitted values and residuals are output. The subsequent three pages relate to a SAS program for fitting Eqn. 5.4 to the data of Lanigan (1963), again showing both the code and the output. The output is identical in form and content to that associated with the fitting of Eqn. 5.1.

Computer code to fit Eqn. 5.1 to the generation time data of Gibson *et al.* (1988) is presented here. The 'input' statement assumes that there are four variables, *viz.* water activity (a_w), temperature (*temp*) in °C, time for a thousand-fold increase in numbers of salmonellae grown in Tryptone Soya Broth (*time*), and generation time of salmonellae (*gentime*). Of the two 'time' response variables, only *gentime* is used in this illustration.

```
data gibson;
title 'combined effect of temperature and water activity';
input aw temp time gentime;
rootrgen=sqrt(1/gentime);
T=temp + 273.16;
cards;
.991 10 176 8.9
.988 10 237 17.7
.980 10 249 16.2
.974 10 288 17.8
.969 10 394 18.4
.991 15  37  2.7
.988 15  41  2.6
.980 15  47  3.2
.974 15  85  6.6
.969 15  98  4.2
.991 20  14  1.0
.988 20  19  1.0
.980 20  20  1.2
.969 20  36  1.9
.991 25 8.5  0.6
.988 25 10.7 0.6
.972 25  16  0.9
.969 25  20  1.0
.988 30 9.6  0.3
.969 30 14.2 0.8
proc nlin method=gauss;
parms d= 0.2
      awmin= 0.95
      Tmin=270.0;
bounds d>0;
bounds awmin<0.969;
bounds Tmin<285;
cons1=sqrt(aw-awmin);
cons2=T-Tmin;
model rootrgen=d * cons1 * cons2;
der.d=cons1 * cons2;
der.awmin=-d * cons2/(2.0 * cons1);
der.Tmin=-d * cons1;
output p=sqrhat r=resid;
proc print;
   var aw T gentime rootrgen sqrhat resid;
run;
```

Non-Linear Least Squares Iterative Phase

Dependent Variable ROOTRGEN Method: Gauss-Newton

Iter	D	AWMIN	TMIN	Sum of Squares
0	0.200000	0.950000	270.000000	0.779370
1	0.280883	0.949410	278.555467	0.585399
2	0.362130	0.949394	279.593597	0.069390
3	0.362689	0.949490	279.361171	0.064988
4	0.362559	0.949467	279.361220	0.064988
5	0.362557	0.949467	279.361171	0.064988

NOTE: Convergence criterion met.

Non-Linear Least Squares Summary Statistics Dependent Variable ROOTRGEN

Source	DF	Sum of Squares	Mean Square
Regression	3	15.120647502	5.040215834
Residual	17	0.064988266	0.003822839
Uncorrected Total	20	15.185635767	
(Corrected Total)	19	3.672107852	

Parameter	Estimate	Asymptotic Std. Error	Asymptotic 95 % Confidence Interval Lower	Upper
D	0.3625571	0.02174056735	0.31668877	0.40842540
AWMIN	0.9494673	0.00295681030	0.94322900	0.95570558
TMIN	279.3611714	0.45501560276	278.40117803	280.32116473

Asymptotic Correlation Matrix

Corr	D	AWMIN	TMIN
D	1	0.8309944324	0.5284485793
AWMIN	0.8309944324	1	0.0481343271
TMIN	0.5284485793	0.0481343271	1

OBS	AW	T	GENTIME	ROOTRGEN	SQRHAT	RESID
1	0.991	283.16	8.9	0.33520	0.28069	0.05451
2	0.988	283.16	17.7	0.23769	0.27036	-0.03267
3	0.980	283.16	16.2	0.24845	0.24066	0.00779
4	0.974	283.16	17.8	0.23702	0.21572	0.02130
5	0.969	283.16	18.4	0.23313	0.19249	0.04064
6	0.991	288.16	2.7	0.60858	0.65012	-0.04154
7	0.988	288.16	2.6	0.62017	0.62620	-0.00603
8	0.980	288.16	3.2	0.55902	0.55742	0.00160
9	0.974	288.16	6.6	0.38925	0.49966	-0.11041
10	0.969	288.16	4.2	0.48795	0.44584	0.04211
11	0.991	293.16	1.0	1.00000	1.01956	-0.01956
12	0.988	293.16	1.0	1.00000	0.98205	0.01795
13	0.980	293.16	1.2	0.91287	0.87418	0.03869
14	0.969	293.16	1.9	0.72548	0.69920	0.02628
15	0.991	298.16	0.6	1.29099	1.38900	-0.09801
16	0.988	298.16	0.6	1.29099	1.33789	-0.04690
17	0.972	298.16	0.9	1.05409	1.02309	0.03100
18	0.969	298.16	1.0	1.00000	0.95255	0.04745
19	0.988	303.16	0.3	1.82574	1.69374	0.13200
20	0.969	303.16	0.8	1.11803	1.20591	-0.08787

Computer code to fit Eqn. 5.4 to the reaction rate data of Lanigan (1963) is presented here. The 'input' statement assumes that there are three variables, *viz.* water activity (a_w), temperature in K (T) and rate of reaction (k).

```
data lanigan;
title 'combined effect of temperature and water activity';
input aw T k;
rootr=sqrt(k);
cards;
.998 297 .5369
.998 303 .862
.998 309 1.1796
.998 315 1.008
.98 297 .4008
.98 303 .5822
.98 309 .794
.98 315 .741
.962 297 .2798
.962 303 .3932
.962 309 .5293
.962 315 .3403
.944 297 .0756
.944 303 .1059
.944 309 .1059
.944 315 .0454
proc nlin method=gauss;
parms d= 0.1
      awmin= 0.94
      Tmin=270
      c=0.3
      Tmax=320;
bounds d>0;
bounds awmin<0.944;
bounds Tmin<297;
bounds c>0;
bounds Tmax<325;
cons1=sqrt(aw-awmin);
cons2=T-Tmin;
cons3=T-Tmax;
cons4=exp(c*cons3);
cons5=1.0-cons4;
model rootr=d*cons1*cons2*cons5;
der.d=cons1*cons2*cons5;
der.awmin=-d*cons2*cons5/(2.0*cons1);
der.Tmin=-d*cons1*cons5;
der.c=-d*cons1*cons2*cons3*cons4;
der.Tmax=d*cons1*cons2*cons4*c;
output p=sqrhat r=resid;
proc print;
  var aw T k rootr sqrhat resid;
run;
```

Non-Linear Least Squares Iterative Phase
Dependent Variable ROOTR Method: Gauss-Newton

Iter	D	AWMIN	TMIN	Sum of Squares
	C	TMAX		
0	0.100000	0.940000	270.000000	0.253423
	0.300000	320.000000		
1	0.120594	0.938441	271.074439	0.027245
	0.384555	318.340152		
2	0.120836	0.938461	270.958070	0.021504
	0.369371	318.810225		
3	0.120938	0.938505	270.962573	0.021444
	0.367787	318.868574		
4	0.120950	0.938511	270.963682	0.021444
	0.367869	318.868878		
5	0.120951	0.938511	270.963826	0.021444
	0.367882	318.868813		
6	0.120951	0.938511	270.963832	0.021444
	0.367883	318.868810		

NOTE: Convergence criterion met.

Non-Linear Least Squares Summary Statistics Dependent Variable ROOTR

Source	DF	Sum of Squares	Mean Square
Regression	5	7.9584558226	1.5916911645
Residual	11	0.0214441774	0.0019494707
Uncorrected Total	16	7.9799000000	
(Corrected Total)	15	1.0826825072	

Parameter	Estimate	Asymptotic Std. Error	Asymptotic 95 % Confidence Interval Lower	Upper
D	0.1209506	0.0388075313	0.03553544	0.20636569
AWMIN	0.9385114	0.0009450743	0.93643131	0.94059152
TMIN	270.9638320	8.9527005468	251.25899566	290.66866826
C	0.3678830	0.5326885824	-0.80456115	1.54032714
TMAX	318.8688102	4.6688348814	308.59273470	329.14488576

Asymptotic Correlation Matrix

Corr	D	AWMIN	TMIN	C	TMAX
D	1	0.04533118	0.99497401	-0.9115604	0.87776880
AWMIN	0.04533118	1	0	0	0
TMIN	0.99497401	0	1	-0.8850271	0.8495211
C	-0.9115604	0	-0.8850271	1	-0.9954279
TMAX	0.87776880	0	0.8495211	-0.9954279	1

OBS	AW	T	K	ROOTR	SQRHAT	RESID
1	0.998	297	0.5369	0.73273	0.76783	-0.035091
2	0.998	303	0.8620	0.92844	0.94232	-0.013878
3	0.998	309	1.1796	1.08609	1.09234	-0.006244
4	0.998	315	1.0080	1.00399	0.98610	0.017897
5	0.980	297	0.4008	0.63309	0.64122	-0.008137
6	0.980	303	0.5822	0.76302	0.78695	-0.023926
7	0.980	309	0.7940	0.89107	0.91223	-0.021163
8	0.980	315	0.7410	0.86081	0.82351	0.037308
9	0.962	297	0.2798	0.52896	0.48247	0.046487
10	0.962	303	0.3932	0.62706	0.59212	0.034938
11	0.962	309	0.5293	0.72753	0.68639	0.041144
12	0.962	315	0.3403	0.58335	0.61963	-0.036275
13	0.944	297	0.0756	0.27495	0.23323	0.041729
14	0.944	303	0.1059	0.32542	0.28623	0.039196
15	0.944	309	0.1059	0.32542	0.33180	-0.006373
16	0.944	315	0.0454	0.21307	0.29952	-0.086452

6.

PREDICTING THE SHELF LIFE OF CHILL STORED FOODS BY TEMPERATURE FUNCTION INTEGRATION

Attempts to predict the remaining shelf life of poultry, fish, milk, dairy products and meat have been made on the basis of the integrated temperature history of the products. For each commodity, a model was developed to describe the effect of temperature on the significant spoilage microbiota, and the shelf life was estimated using devices such as temperature integrators or loggers programmed with the appropriate model.

6.1. SPECIFIED SPOILAGE LEVELS

Spencer and Baines (1964) introduced the term "specified critical level", e.g. a flavour score of 4 on the numerical scoring system of Shewan *et al.* (1953) for wet fish. The term was altered by Olley and Ratkowsky (1973a, b) to "specified spoilage level" which represents the time necessary to reach a certain point in a sequence of events. As foreshadowed by Arrhenius (1915), this is the basis from which the concept of relative spoilage rates at different temperatures was developed and is a consequence of the balanced growth exhibited by bacteria (see 2.3.1. and 3.1.). Thus bacteria growing on a particular substrate produce the same metabolite over a wide temperature range and there is a direct proportionality between the number of cells present and the level of substrate utilised or product formed, e.g. Sasajima (1973, 1974) found that bacteria isolated from fish produced a spoilage level of 16.2 μg/mL trimethylamine (TMA) from trimethylamine oxide (TMAO) in a broth culture, when the cell density was $10^{7.5}$-10^{8} CFU ml^{-1} across the temperature range -4°C to 25°C. The results of Sasajima are in accord with the deterministic model of bacterial growth envisaged by Jason (1983) and discussed in 2.3.3.2.

As soon as bacteria have multiplied sufficiently to be counted accurately, or have produced enough metabolite to be measured, the specified spoilage level can be selected. If a metabolite is chosen it must, of course, be an end-product of metabolism and not a compound further metabolised, in which case the proportionality between microbial numbers and end-product concentration would not hold.

The level of metabolite chosen will obviously be less than that causing overt spoilage to allow prediction rather than diagnosis of spoilage, e.g. Codex Alimentarius specifies 10-15 mg of trimethylamine nitrogen (TMA-N) per 100 g of fish muscle as unacceptable, but the level 1.5 mg recommended by Connell (1975) for very good quality cod could equally well be selected.

Once the relationship between storage conditions and microbial growth or metabolite production has been established, the level can be specified as any combination of time and temperature (or other rate controlling factor). This eliminates the need for bacterial enumeration and allows the specified level to be set at an early stage of deterioration where the amount of metabolite produced may be below the level of detection even by sophisticated chemical instrumentation.

For food-borne pathogens which produce infections or intoxications, appropriate levels are those which take account of an infectious dose or level of toxin required to cause disease. For dangerous pathogens, a nil tolerance may be required and conditions of product formulation and storage recommended to prevent resolution of lag phase. With indicator organisms such as *Escherichia coli*, a cut-off point may be selected to represent the level at which a process is considered unhygienic (Gill *et al.* 1988b). Temperature integration to assess process hygiene will be considered in Chapter 7.

6.2. THE RELATIVE RATE CONCEPT

The introduction of the relative rate concept represented an important advance in the application of time/temperature spoilage relationships to predict shelf life. An early attempt to derive a suitable relative rate function was described by Nixon (1971) at a seminar on fish quality held in Wellington, New Zealand. The relative spoilage rate was calculated as:

$$\text{Relative rate of spoilage at temperature } t = \frac{\text{keeping time at } 0°\text{C}}{\text{keeping time at temperature } t} \tag{6.1}$$

This enabled integration of temperature history to give an equivalent time on ice (Table 6.1).

The linear relative spoilage rate proposed by Nixon (1971) was based on limited data sets and was refined by Olley and Ratkowsky (1973a, b), as mentioned in Chapter 1, who were encouraged in their quest for a 'universal' spoilage model by observations that a variety of deteriorative processes could be described by Arrhenius kinetics with similar activation energies, e.g. spoilage of various species of shark and mortality of abalone (Olley, 1971). The OR curve (Olley and Ratkowsky, 1973a, b) was based on 70 data sets from the literature and included specified spoilage levels based on bacteriological, chemical, physical and organoleptic criteria. In turn, this gave rise to the square root model, Eqn. 3.10 (Ratkowsky *et al.*, 1982), in which a T_{min} value of 263K gives a relative rate function that appears to provide a reasonable

Table 6.1 **Temperature History Profile for 'On Board' Fish Chilling Process Converted to Equivalent Times in Ice by a Relative Rate Function**

Time	Temperature	Relative Spoilage Rate		Equivalent Hours in Ice-Rate x Time	
(h)	(°C)	Nixon	OR†	Nixon	OR†
2	15.0	7.0	6.5	14.0	13.0
1	14.7	6.9	6.4	6.9	6.4
1	13.5	6.4	5.8	6.4	5.8
1	10.8	5.3	4.3	5.3	4.3
1	9.4	4.8	3.6	4.8	3.6
4	8.6	4.4	3.3	17.6	13.2
2	7.5	4.0	2.9	8.0	5.8
2	5.0	3.0	2.7	6.0	5.4
10	3.5	2.4	1.8	24.0	18.0
24				**93.0**	**75.4**

OR† : Olley/Ratkowsky relative rate function.

[Adapted from Nixon (1971)].

description of the effect of temperature on the spoilage rate of a wide range of proteinaceous foods stored aerobically at chill temperatures where pseudomonads are the dominant spoilage microbiota. As the amount of information on a particular product increases, however, different T_{min} values, and as a result, a different relative rate function may be more appropriate. The effect on relative rate of changing T_{min} from 263K to 264.5K is shown in Table 6.2, opposite.

There is increasing evidence that the latter T_{min} may be more appropriate for spoilage of raw or pasteurised milk (Chandler and McMeekin, 1985a, b; Griffiths and Phillips, 1988a; see 6.4.3.) and minced beef (unpublished results). An additional column has been included in Table 6.2 in which the reference value of 1.00 has been set at 4°C, the recommended storage temperature for milk and dairy products.

In addition to introducing the relative rate concept, Nixon (1971) developed the prototype of a solid state temperature integrator, subsequently patented by Nixon in New Zealand in 1975 and in the USA in 1977 as the TEFIMUPOT (see 9.4.1.). The circuitry incorporated the OR curve and the device was designed to enable the use of multiple sensors in fish holds with telemetry delivering the information to a central control station.

Gill (1986) perceived some major problems that might arise in the application of temperature function integration to spoilage prediction. Assuming that a suitable model is available (probably based on Bělehrádek or Arrhenius kinetics), the problems he foresaw were:

i) varying initial numbers of contaminants, and
ii) the complexity of the systems being analysed in that the model must encompass the temperature dependence of all species in the spoilage biota. In addition, the fraction of the initial biota that will grow will vary with changing temperature.

These considerations led Gill (1986) to the conclusion that: *"it is inherently unlikely that any single equation could model such a complex situation. Temperature function integration therefore may have very restricted uses for spoilage prediction"*.

Nevertheless, despite the apparent difficulties, several investigators encouraged by the simplicity of the square root model, were prepared to consider the matter further. Their conclusions are considered in the remainder of this chapter.

Table 6.2 **Relative Spoilage Rate Table for Proteinaceous Foods Based on the Square Root Model (Eqn. 3.10)**

Temperature (°C)	Rate of spoilage relative to: 0°C		4°C
where T_{min} = :	263K	264.5K	264.5K
-1	0.81	0.78	0.36
0	1.00	1.00	0.46
1	1.21	1.25	0.58
2	1.44	1.53	0.71
3	1.69	1.83	0.85
4	1.96	2.16	1.00
5	2.25	2.52	1.17
6	2.56	2.91	1.35
7	2.89	3.32	1.54
8	3.24	3.77	1.74
9	3.61	4.24	1.96
10	4.00	4.74	2.19
11	4.41	5.26	2.43
12	4.84	5.82	2.69
13	5.29	6.40	2.96
14	5.76	7.01	3.24
15	6.25	7.64	3.53

* T_{min} taken as -10°C; † T_{min} taken as -8.5°C

6.3. TIME-TEMPERATURE TOLERANCE (TTT)

The TTT hypothesis is the third of the concepts underlying the application of temperature function integration to predict microbial behaviour in foods. The concept was presented succinctly in the I-POINT product manual (see 8.3.1.) and is reproduced verbatim in 6.3.1. to 6.3.4., below. The basic principle of TTT applies to all devices, however, and these are considered in detail in Chapters 8 and 9.

6.3.1. Prerequisites

"Chill preservation is probably the best existing method to lengthen the quality life of perishable commodities, such as food and pharmaceutical products. To take the full advantage of this preservation method, however, there are some basic prerequisites to carefully consider.

i) the initial quality of the product subject to chill preservation must be of an acceptable quality
ii) the temperature handling throughout the whole distribution chain must be properly taken care of, otherwise all good intentions in the first step will prove to be meaningless

6.3.2. Deterioration Processes

"The quality deterioration of a perishable product is mostly caused by the activity of a number of breakdown processes, most of them time-temperature dependent. The speed of their reactions depends upon temperature, meaning that:

i) the higher the temperature the higher the speed[1]
ii) the lower the temperature the lower the speed.

6.3.3. The TTT Hypothesis

"This is also in essence what forms the basis for the fundamental rule of chill preservation: the so called Time Temperature Tolerance hypothesis, which in conclusion states:

i) quality loss is proportional to the integrated accumulated effects of both time and temperature;
ii) the rate of quality loss will increase as storage temperature increases, and decrease as storage temperature decreases;
iii) all quality losses are cumulative and irreversible;
iv) all temperature effects are commutative, i.e. they have the same result on product quality no matter when they occur during the life of the product.

6.3.4. TTT Data for Products

"Thus, what is needed to calculate true TTT data for a perishable product is simply knowledge of the characteristics of those quality deteriorating processes that affect the product in question. Such facts exist to a great

---

[1] In the suboptimal range

extent already and also have to be considered when a product is subject to date marking".

6.4. VALIDATION AND PRACTICAL APPLICATION OF TEMPERATURE INTEGRATION

6.4.1. Poultry Spoilage

Daud *et al.* (1978) applied the OR curve to the rate of spoilage of chicken tissues, the development of spoilage bacteria and the utilisation of amino acids as a function of temperature. They concluded that the remaining shelf life of poultry tissues may be predicted at temperatures up to 16°C by using a temperature function integrator programmed with the OR curve. In the region 16 - 25°C relative rates continued to increase with temperature and did not plateau at a relative rate of ~9 as predicted by the OR curve. The continued increase in relative rate from 16 - 25°C would be expected given that the optimum temperature of psychrotrophic pseudomonads is 25 - 30°C.

Pooni and Mead (1984) applied the models of Spencer and Baines (1964), Olley and Ratkowsky (1973b) and the square root model to 14 studies incorporating 28 data sets on spoilage of chicken, turkey and duck. They noted that, in poultry meat stored aerobically, factors other than storage temperature affect the *absolute* spoilage rate but not the *relative* rates of spoilage at different temperatures.

For example, red and white muscle differ in chemical composition and pH value which may affect the behaviour of the spoilage biota. Despite these differences, pseudomonads always dominated at spoilage and the substrate effect was concluded to be much less than the effect of temperature on spoilage rate. As the actual time to spoilage is directly related to the initial level of psychrotrophic contamination (Barnes *et al.*, 1978), the application of a rigorous quality assurance program will reduce variation in product contamination, ensure more uniform shelf life, and enable more confident application of temperature function integration.

6.4.2. Fish Spoilage

The Olley and Ratkowsky (1973a) spoilage curve which led to the rediscovery of the Bělehrádek equation was based largely on information pertaining to fish spoilage. From the data sets used, a T_{min} value of 263K appears to describe an appropriate relative rate curve for psychrotrophic spoilage. As noted in 6.2., however, for particular products and fish harvested in different geographical locations, it may be worthwhile to determine T_{min} values of the dominant spoilage microbiota at chill temperatures.

An attempt to verify the use of the OR spoilage curve for tropical fish transhipped to the UK on ice was reported by Olley and Quarmby (1981). In this work, storage life at 5°C and 10°C was predicted on the basis of storage life at 0°C using taste panel and chemical assessment of deterioration. The T_{min} value chosen (263K) overpredicted for one species and underpredicted for the other, but predictions for both were within acceptable limits. Subsequently, a T_{min} of 263K was also found to be suitable for predicting the storage life of common carp in Israel (Gelman *et al.*, 1990).

The relationship also appears to hold for fish packaged in modified atmospheres. Gibson (1985) combined organoleptic assessment and conductance assays to predict the shelf life of modified atmosphere packed and vacuum-packed fish. Using the relative rate function calculated from the square root model with a T_{min} of 263K, the theoretical shelf lives of fish stored at 5°C and 10°C, based on the rate at 0°C, were calculated (Table 6.3, opposite).

Agreement between experimental and calculated shelf lives was very good and bacteriological counts on samples taken at the times indicated were very similar (1.3-6.7 x 10^7/g) regardless of storage temperature or the time the pack had been stored. Similarly, the conductance detection times corresponding to a cooked flavour score of 5.5 were virtually the same regardless of storage time and temperature.

It was estimated that, when the cooked flavour score reached the specified spoilage level [5.5 on the scale developed by Shewan *et al.* (1953)], the bacterial count would be 3.1 x 10^7/g and the conductance detection time 10 h. As cooked flavour score is linearly related to the storage time of fish on ice, and there is a good correlation between cooked flavour score and detection time, it was possible to calculate the change in the quality parameters over the total time of the experiment and hence the daily rate of change for each quality parameter, as follows:

$$\frac{P - P_o}{dP/dt_{\,temp}} = \textit{shelf life} \text{ (in days)}$$

Table 6.3 **Time Since Packaging (Shelf Life) for Packaged Cod to Reach Cooked Flavour Score = 5.5 Units, Predicted Using Relative Rate Function.**

Storage Temperature (°C)	Days to Cooked Flavour Score = 5.5 Units			
	Modified Atmosphere Packs		Vacuum Packs	
	Experimental	Calculated	Experimental	Calculated
0	14		10	
5	6	6.2	4	4.4
10	3	3.5	2	2.5

Table 6.4 **Daily Change of Quality Parameter (dP/dt) for Modified Atmosphere Packs and Vacuum Packs.**

Storage Temperature (°C)	Modified Atmosphere Packs [Detection times (h)]		Cooked Flavour (Units)	Vacuum Packs [Detection times (h)]		Cooked Flavour (Units)
	Sea Water Broth	Wood and Baird's Broth		Sea Water Broth	Wood and Baird's Broth	
0	0.9	0.61	0.25	0.96	0.55	0.3
5	2.0	1.37	0.6	2.2	1.24	0.7
10	3.6	2.44	1.0	3.8	2.75	1.2

Tables 6.3 and 6.4 *after* Gibson (1985) with permission of Blackwell Scientific Publications, Ltd.

where:

P is the value of the parameter measured

P_o is the value of the parameter at the cut-off point and

dP/dt_{temp} is the spoilage rate shown in Table 6.4.

Gibson's verification of the usefulness of the 263K predictive relationship was supported by Scott *et al.* (1986) who noted that the shelf life of CO_2 -flush packs of snapper at -1°C was close to that predicted from previous experiments at 3°C.

Two constraints on the model noted by Gibson (1985) were:

1) that it was useful only when spoilage was mainly bacterial but could not be extended to processes such as chemical rancidity, and
2) that as the model was developed at constant temperature it should be applied strictly only to products stored at constant temperature until tested under fluctuating temperature regimes.

In fact, reassessment of work carried out at Torry Research Station more than 20 years earlier (Spencer and Baines, *unpublished data*) indicated that temperature function integration worked in practice whether fish had been stored at 1°C, 15°C or cycled between these temperatures for periods equivalent to equal number of days at 0°C (Fig. 6.1, opposite).

Gibson's first criticism was addressed by Bremner *et al.* (1987) who proposed the use of a rapid systematic sensory method to assess the quality of seafoods based on a custom-designed electronic calculator incorporating the demerit point scheme of Bremner (1985). The sensory scheme (and hence the calculator) operated as a temperature function integrator as the authors demonstrated that most of the major deteriorative influences contributing to deterioration of seafood in storage were described by the square root model (Table 6.5).

Importantly, most of the major changes had similar T_{min} values indicating that each could be described by the same relative rate function. The obvious exception is the physical change in dielectric properties measured by the Torrymeter. This had a T_{min} value of 250K and would respond to temperature in a different manner than that appropriate for microbial or biochemical changes.

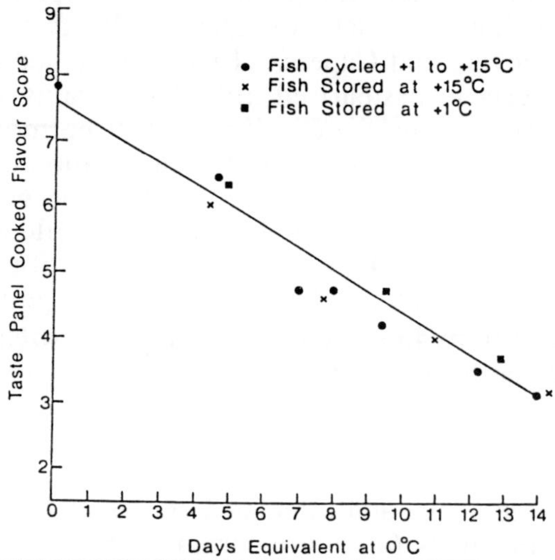

Fig. 6.1 Taste panel flavour scores of cod *versus* equivalent days on ice. Fish cycled at 1°C to 15°C, stored at 15°C and 1°C. [Reproduced from McMeekin *et al.* (1988) by permission of CRC Press, Inc.].

Table 6.5 **Major Deteriorative Influences of Seafood Described by the Square Root Relationship**

Category	T_{min} (K)	Reference
Psychrotrophic microbial growth	263-266	Ratkowsky *et al.* (1982)
Nucleotide breakdown (2 - 25°C)	265	Ohta and Hirahara (1977)
Nucleotide breakdown (-10 - +30°C)	261	Miki and Nishimoto (1984)
Myoglobin breakdown	263	Miki and Hamada (1977)
		Miki and Nishimoto (1984)
Autolytic changes	258 - 261	Gildberg (1982)
Dielectric properties	250	Jason and Richards (1975)
Bulk Properties	264	Barlow (1976)

6.4.3. Milk and Dairy Product Spoilage

Perhaps greatest interest in shelf life prediction by temperature function integration has been shown in relation to short shelf life dairy products. In this context, Langeveld and Cuperus (1980) researched the effect of temperature on bacteria in pasteurised milk. The strains studied were representative of thermoduric organisms and thermolabile post-pasteurisation contaminants important in the deterioration of pasteurised milk at different temperatures. A computer programme was developed and tested to determine if bacterial numbers in milk could be calculated from the Arrhenius equation and a given temperature profile.

Activation energies calculated from the *central* portion of each Arrhenius plot ranged from 45 - 110 kJ mol^{-1}. For *Enterobacter hafniae* B2, the Arrhenius equation was given by:

$$ln\ k = C - \mu / RT = 39.6 - 975/RT$$

A comparison of the actual and calculated counts of *E. hafniae* B2 in milk at storage temperatures *fluctuating* from 7-25-7-25-7°C is shown in Fig. 6.2. Clearly, there was good agreement between actual and calculated counts.

A further important result in the paper of Langeveld and Cuperus (1980) was the finding that when strains were grown together in milk the growth rates were essentially the same as when they were grown separately in the same substrate. For *Streptococcus lactis,* generation times at 27°C were 0.67 h and 0.64 h in single and mixed culture; for *Bacillus cereus* 0.66 h and 0.62 h, and for a *Pseudomonas* strain 0.73 h and 0.77 h.

The doubling times of the bacterial population in pasteurised milk estimated by Langeveld and Cuperus (1980) have been used to compare Arrhenius and square root plots (Fig. 6.3). Two lines of different slope have been drawn in Fig. 6.3a to exemplify the ease with which a continually downward sloping curve may be interpreted as a biphasic plot (see also Fig. 2.2).

Obviously, the square root model provides a better description of the experimental results, a point noted by Langeveld (1983) in a subsequent communication. Nevertheless, from a practical viewpoint, he was of the opinion that the predictions obtained would be unaffected by the model used.

Chandler and McMeekin (1985a, b) applied the square root model to data from the literature on the growth of bacteria of dairy origin in laboratory media and milk and obtained a good description of

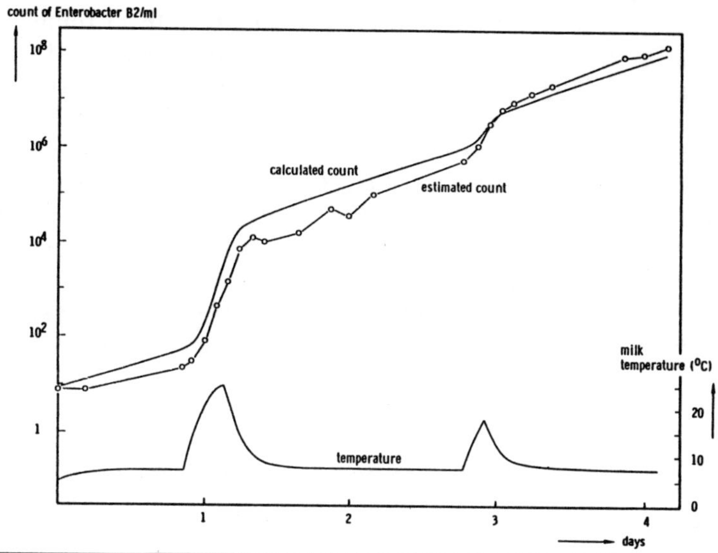

Fig. 6.2 Growth of *Enterobacter hafniae* B2 in milk at fluctuating temperatures, comparing calculated and estimated counts. [Reproduced from Langeveld and Cuperus (1980) by permission of Centre for Agricultural Publishing and Documentation, Wageningen, The Netherlands].

experimental results. Similarly, the authors' experimental data for raw and pasteurised milk confirmed the validity of the square root model. Based on time to reach a value of log 7.5 cfu ml^{-1}, T_{min} values of 264.5 and 264.8K were estimated for raw milk and a mean value of 264.7K was obtained for five samples of pasteurised, homogenised milk. The values were remarkably similar to the T_{min} estimate of 264.5K obtained from a square root plot using the data of Langeveld and Cuperus (1980) for pasteurised milk (Fig. 6.3). The average T_{min} value obtained by Griffiths and Phillips (1988a) for 46 samples of raw milk stored between 1.5°C and 15°C was 264.4K.

An important aspect of the work of Chandler and McMeekin (1985b) was to examine the response of lag time to temperature. Using stationary phase cells of *Pseudomonas* strain E5.2, a T_{min} value of 267.3K was estimated. This compared with 267.9K for growth rate data, indicating that both the lag and logarithmic phases of growth can be described by the same relative rate function. Similarly, Fu *et al.* (1991) found almost identical T_{min} values for the temperature dependence of the lag (265.5K) and log (265.3K) phases of growth of the important dairy spoilage organism *Pseudomonas fragi.*

The relation between temperature and growth of individual bacteria in dairy products was investigated in detail by Phillips and Griffiths (1987). Fifteen Gram-negative strains and 14 Gram-positive strains isolated from dairy products were grown in skimmed milk, full-fat milk, single cream and double cream at a variety of temperatures. Square root and Arrhenius approaches were applied to the growth rate results and equations derived to relate various growth parameters to the temperature of growth. T_{min} values were found to be similar in different products and although estimates of activation energy obtained from the Arrhenius equation were more variable, a correlation between T_{min} and activation energy was noted. This, of course, follows from Eqn. 3.24:

$$\Delta E_{app} = 2RT^2/(T - T_{min}).$$

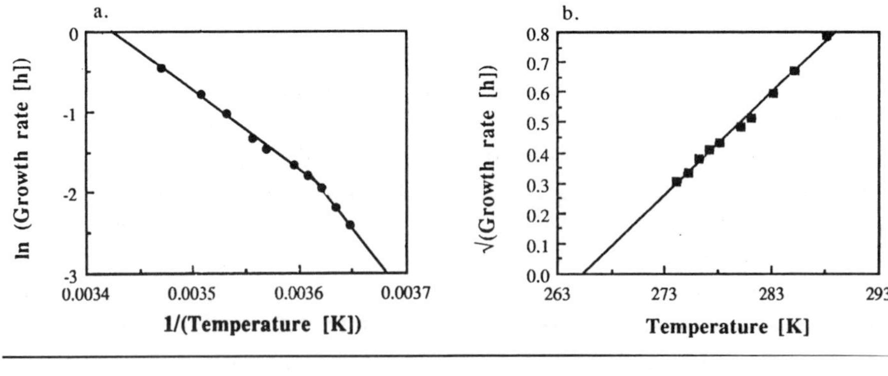

Fig. 6.3 Doubling times of bacteria in milk estimated by Langeveld and Cuperus (1980) and presented as: a) Arrhenius, and b) square-root plots.

In addition, T_{min} values in mixed culture were generally the same as in axenic culture, confirming the report of Stannard *et al.* (1985) for bacteria grown singly and in various combinations in laboratory media.

Phillips and Griffiths (1987) also used various growth parameters to determine T_{min} values. These were the square roots of the reciprocal of lag phase, time to maximum growth rate, time to stationary phase, time for a 4-log cycle increase and growth rate. The authors reported remarkably good agreement between T_{min} values estimated by the different criteria (Fig. 6.4).

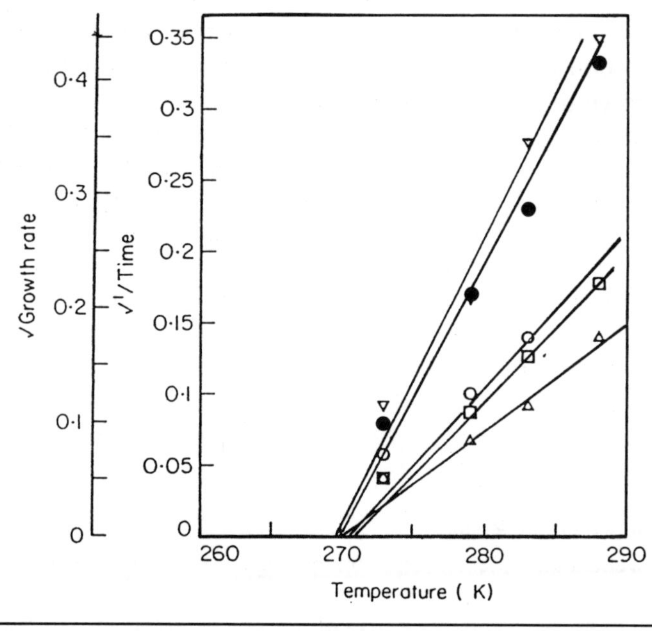

Fig. 6.4 Square-root plots of *Serratia marcescens* 755 growth in full fat milk using different criteria: length of lag phase (▽); time to maximum growth rate (O); time to stationary phase (Δ); time for a 4-log cycle increase in population size (□); and growth rate (●). [Reproduced from Phillips and Griffiths (1987) by permission of Academic Press Inc.].

Griffiths *et al.* (1987), in a study of the effect of extended low temperature storage of raw milk on the quality of pasteurised and UHT milk found that, on average, milks stored at 2°C lasted 7.4 days before the bacterial count reached 1 x 10^7CFU ml^{-1}, the number at which appreciable levels of proteases and lipases are synthesised. At 6°C this level was reached in 4.1 days, i.e. 1.8 times faster than at 2°C. From Table 6.2 the relative rate predicted between 6°C and 2°C is 1.78 (=2.56/1.44) for a T_{min} value of 263K, and 1.9 (=1.35/0.71) for a T_{min} of 264.5K.

Griffiths and Phillips (1988 a, b) continued their modelling studies to consider milks of varying hygienic quality. Whilst the effect of temperature was accurately described by square root plots, T_{min} values of the naturally occurring microbiota varied considerably and appeared to be related to the quality of the pasteurised milk. Gram-positive bacteria, which had higher T_{min} values, were isolated in greater numbers from good quality milks and pasteurised milks with high T_{min} values were shown to have longer shelf lives.

As the constant *b* in the square root equation (Eqn. 3.10) was relatively unchanged between pasteurised milk samples, Griffiths and Phillips (1988b) argued that an effective means to predict T_{min} values would allow prediction of shelf life at any storage temperature. This was achieved by counts obtained on various media after preincubation of the sample for 25 h at 15°C, e.g. a count on milk agar + Oxoid CFC antibiotic supplement was correlated with the T_{min} value (n = 46; r = 0.73; p <0.001) giving rise to the predictive equation:

$$\text{shelf life (h)} = 1/[0.0621(T - (269.55 - 0.74(CFC_{15}) - 0.11(CFC_{15})^2]^2$$

where:

T = storage temperature in K, and

CFC_{15} = $\log_{10}$ count after preincubation on milk agar + CFC supplement.

The accuracy of this type of predictive equation increased with increasing storage temperature, mainly because of the range of shelf lives at 14°C (1.4-6.1 days) is smaller that that at 2°C (3.25-138 days). This reflects a greater variance of shelf life at low temperatures, where spoilage rates are low, than at temperatures where spoilage rates are high (see also 4.4. and 7.2.).

The approach taken by Chandler and McMeekin (1989c) to predict shelf life of pasteurised, homogenised milk differed from that described above by Griffiths and Phillips (1988b) in which shelf life of individual samples was related to T_{min} values. The essential difference is that the former authors argue that it is valid to consider only the "organisms of concern", i.e. pseudomonads in the case of chill stored milk and dairy products. Underestimating the shelf life of some samples was not thought to be a problem in that if the milk is wholesome by the use-by date the consumer and, as a consequence, the processor, will be satisfied.

Chandler and McMeekin (1989c) used a Time/Temperature Function Integrator (TTFI, Don Whitley, West Yorkshire, United Kingdom), programmed with a relative rate function based on a T_{min} value of 263K and with 4°C as the reference temperature (i.e. relative rate at 4°C=1). TTFI readings (equivalent time elapsed at 4°C) were correlated with bacterial numbers and types and with organoleptic changes in the product during storage.

At temperatures up to 12°C, pseudomonads constituted the entire microbiota of spoiled samples but represented only 10-20% of the biota at 15°C. Because of this result, correlation of TTFI readings with microbial numbers was restricted to a maximum temperature of 12°C. The relationship between pseudomonad numbers and TTFI readings is shown in Fig. 6.5.

The linear relationship for pseudomonads extends from levels representing initial contamination (10 CFU ml^{-1}) to those at which the product would be spoiled ($10^{7.5}$ CFU ml^{-1}) suggesting that TTFI can be used directly to monitor elapsed and remaining shelf life.

It was recognised that time to reach a specified spoilage level will depend on the initial level of contamination of the product, a point raised by Pooni and Mead (1984), Gill (1986) and others. Whilst these authors viewed this as a serious difficulty in the application of TTFI to shelf life prediction, Chandler and McMeekin (1989c) argued that the problem may be overcome by specifying a maximum initial level of contamination for the organisms of concern. The level is set by that obtainable by good manufacturing practice and takes account of the shelf life required for a particular product. For pasteurised, homogenised milk in Australia, an initial level of 10 pseudomonad-CFU ml^{-1} and a shelf life of 8 days was considered appropriate.

A practical benefit of TTFI in the dairy industry is that it may be used to identify whether unsatisfactory processing procedures, under the processor's control, or subsequent temperature abuse beyond the processor's control, is responsible for poor keeping quality. In a storage trial at the recommended temperature, if the product has spoiled before the expected shelf life is reached, poor hygiene conditions are indicated.

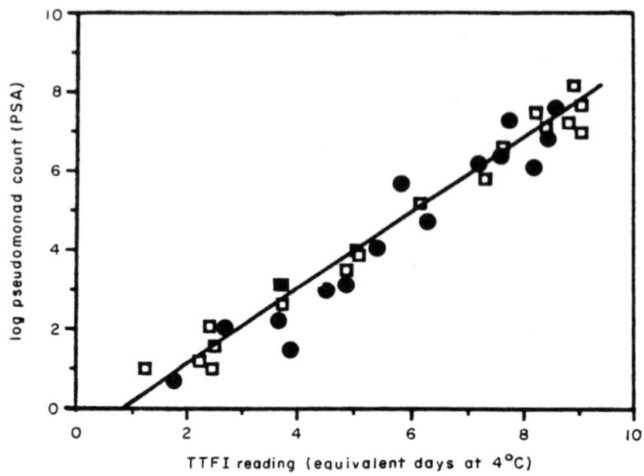

Fig. 6.5 Correlation of *Pseudomonas* numbers with TTFI readings from pasteurised, homogenised milk stored at 5 - 6°C (□) and 10 - 12°C (●). [Reproduced from Chandler and McMeekin (1989c) by permission of Academic Press Inc.].

To obtain a result more rapidly, Chandler and McMeekin (1989c) proposed an accelerated test at 12°C (i.e. maximum temperature at which pseudomonads totally dominate the microbiota). The observed shelf life at 12°C was linked directly to the predicted shelf life at 4°C by application of the appropriate relative rate factor (2.47). If the expected shelf life at 4°C is 8 days, the test at 12°C will be complete in 3.2 (=8/2.47) days.

The practical application of TTFI also extends to evaluation of storage and distribution procedures. A marked advantage is that it does not only identify an endpoint, but continually monitors product deterioration. Also, loss of shelf life at any part of the distribution chain can be identified.

A further application of temperature integration lies in the selection of preincubation conditions commonly used in dairy microbiology to build up bacterial numbers before enumeration by viable counts or other methods. The temperature selected will influence the composition of the microbiota at the time of enumeration and may select for certain types of microorganisms e.g. psychrotrophs or mesophiles.

Maxcy and Liewen (1989) examined the proposition that preincubation of raw milk samples for 18 hours at 12.8°C (55°F) enhanced growth of a segment of the microbiota commonly associated with unsanitary methods and conditions on the farm. Using pure cultures of *P. fluorescens* and *E. coli* as well as enriched mixed cultures of psychrotrophs and mesophiles the logarithm of the generation time was plotted against temperature to estimate the slope. Our evaluation of the results by the square root model gave the expected linear relationship and appropriate T_{min} values for each of the cultures studied. For mesophiles T_{min} was 276.4K and for psychrotrophs 267K. The overlay plots of the fitted square root models for mesophiles and psychrotrophs, shown in Fig. 6.6, indicate that the two populations have equal growth rates at 22°C and that below this temperature the psychrotrophs grow increasingly faster than the mesophiles. At 12.8°C, the generation time predicted for psychrotrophs was 2.21 h and that for the mesophile population was 4.07 h. During an 18 h preincubation period at this temperature the different rates of growth would allow for the production of approximately 10

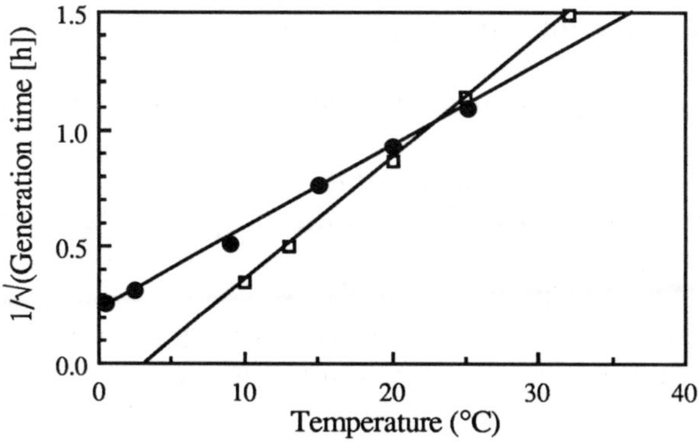

Fig. 6.6 Square-root plots for psychrotrophs (□) and mesophiles (●) in raw milk. [Data of Maxcy and Liewen (1989)].

times the number of psychrotrophic cells as compared with the mesophilic cells, e.g. starting with 10^3 cells of each type, 8.14 generations of psychrotroph growth would occur, but only 4.42 generations of mesophile growth, resulting in populations of ~2.5 x 10^5, and ~2.5 x 10^4 cells, respectively.

The application of growth rate data as above to target organisms or populations combined with a knowledge of their expected initial numbers in the product will allow selection of preincubation conditions to enhance growth of the organisms of concern, compared with other components of the population, i.e. selective enrichment based on temperature responses.

6.4.4. Red Meat Spoilage

The approach adopted by Chandler and McMeekin (1989c) appears to overcome most of the problems cited by Gill (1986) as limiting the application of temperature function integration to shelf life prediction, a view reinforced by Gill *et al.* (1988a). In this paper, the New Zealand group described a computer programme for assessing the storage life of chilled red meats. The work was prompted by the necessity to ensure that chilled New Zealand meat products in international trade were suitable for further processing, distribution and retail sale in the chilled state. This could not be achieved by conventional procedures, but it was argued that the remaining storage life could be assessed confidently from information on the storage characteristics of the product and the temperature history. This information was incorporated into a computer programme that could interrogate a temperature logger to indicate the remaining shelf life.

The variables determining meat storage life were perceived to be:

(i) the initial number of contaminants;
(ii) the composition of the product (particularly pH and carbohydrate content);
(iii) the type of packaging used;
(iv) the temperature history of the product which can be recorded routinely using an electronic logger.

The effects of the variability inherent in (i) and (ii) were overcome by obtaining information on deterioration of the product at a known constant temperature, i.e. knowing the assured storage life at a defined temperature and the number of bacteria required to cause spoilage allowed calculation of a notional value for initial numbers.

The proposed model assumed that meats under anaerobic conditions were spoiled by psychrotrophic enterobacteria and that spoilage occurred at 10^7 CFU cm^{-2}. Models were developed separately for the lag and logarithmic phases of growth under anaerobic conditions and in a CO_2 enriched atmosphere.

During operation, the temperature history is first analysed for periods above 7°C as these may allow growth of pathogens. If the appropriate programme for this function indicates more than three generations of *E. coli* (Gill *et al.* 1988a), a possible health hazard exists. This programme is considered in greater detail in Chapter 7.

For shelf life estimation, the entire temperature history profile was used to estimate the increase in numbers of the psychrotrophic enterobacteria and hence the remaining shelf life. Only limited validation of the programme has been carried out, but the results indicated good agreement between actual and calculated storage life. This suggests that temperature function integration may, despite the reservations expressed by Gill (1986) and cited in 6.2., prove to be a useful procedure in shelf life evaluation. Gill's subsequent enthusiasm for temperature function integration as an effective means to assess the hygienic adequacy of processing procedures will become evident in Chapter 7.

6.5. SUMMARY

Development of the concepts of specified spoilage levels, relative rates and time-temperature tolerance are the pillars upon which successful application of temperature function integration depend. Most studies concerned with model validation have indicated that there is considerable potential for practical application in estimating the remaining shelf life of poultry, fish, milk and dairy products and red meat.

Devices for temperature integration are based largely on relative rate functions derived from a square-root model with a T_{min} value of 263K. There is increasing evidence, however, that using a T_{min} of 265K may provide more accurate shelf life predictions.

To overcome the problems of variable initial numbers of psychrotrophic contaminants, a realistic starting level may be specified and the integrated temperature history required to reach the specified spoilage level may be estimated.

7.

EVALUATION OF HYGIENE AND MICROBIOLOGICAL SAFETY OF PRODUCTS AND PROCESSES

The proposed use of temperature function integration to model growth of psychrotrophic bacteria in the spoilage of chill-stored food prompted several groups of workers to examine the potential of the technique to assess the hygienic adequacy of meat processing procedures. Under these circumstances the organisms of concern are members of the Enterobacteriaceae (coliforms, *Escherichia coli* and *Salmonella*) that display mesophilic temperature characteristics. Studies on the development of appropriate models, and validation in laboratory scale experiments and commercial operations, lead to the conclusion that temperature function integration can predict accurately the growth of mesophilic indicator and pathogenic microorganisms. Thus, laborious and inaccurate plate counting procedures normally used for the direct determination of bacterial growth may be replaced by this technique.

7.1. DEVELOPMENT AND VALIDATION OF MODELS FOR *ESCHERICHIA COLI* AND *SALMONELLA* PROLIFERATION DURING MEAT PROCESSING

7.1.1. Offal Chilling Processes

The first application of temperature function integration in this mode was reported by Gill (1984) in an attempt to monitor early spoilage of livers. In ecological terms, this was a simple situation in which the microbiota was dominated by *Escherichia coli*. At most temperatures, values for growth rates of six strains isolated from livers were within 10% of the average. Wide discrepancies occurred only at the extremes where growth was extremely slow above 44°C and below 8°C. A plot of the square root of growth rate *versus* temperature gave a straight line relationship over much of the range, but a distinct change of slope was reported above 30°C (Fig. 7.1a). This response is at variance with the data of Ingraham *et al.* (1983), Smith (1985) and Mackey and Kerridge (1988) who demonstrated a linear square root response from the minimum temperature at which growth rate was determined to at least 35°C. The reason for the discrepancy is not clear; however, the response may be interpreted as linear as analysis of the original data between 7°C and 35°C indicates that it can be described by a straight line.

Nevertheless, use of the biphasic model to calculate *E. coli* increases under several cooling regimes differed by less than one standard deviation from average values of direct determinations. Both manual and computer aided calculations of growth based on the square root response to temperature gave values within 0.5 generations of each other. This contrasted with the high standard deviations obtained for most data sets obtained by direct determinations. We will comment again later on the inappropriateness of standard deviations as a means of summarising growth rate data. Gill and Harrison (1985) noted that, as variations of an order of magnitude can be expected for replicate plate counts, calculated values probably provide a better practical guide than direct estimation.

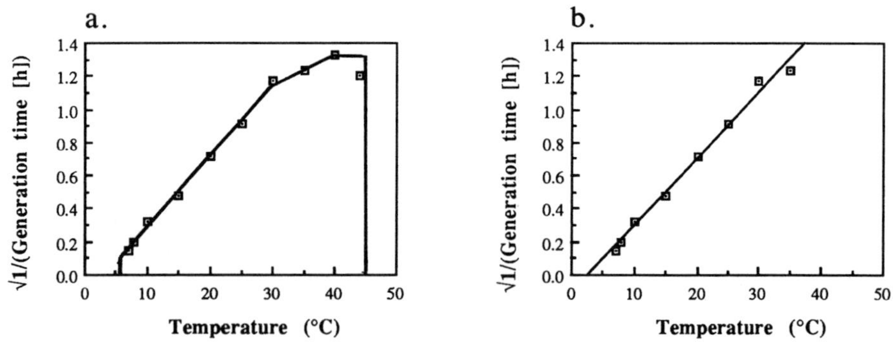

Fig. 7.1 **a**. Square-root plot of growth rate *versus* temperature for anaerobic growth of *E. coli* [*after* Gill and Harrison (1985), with permission of Academic Press, Inc.]. **b.** Square root plot of the same results, using data from 7°C - 35°C only (correlation coefficient = 0.99, T_{min} = 2.76 °C).

To allow application of the technique to any cooling procedure the concept of a Bacterial Growth Number, i.e. the number of generations occurring during a process, was introduced. Provided this number was below a tolerable maximum, the exact form of the cooling profile would be irrelevant, and regulatory and quality assurance procedures would no longer depend on time-consuming and inaccurate plate count procedures.

The results described were applicable only to liver cooling which, as noted earlier, was a relatively simple ecological situation in which an organism of concern was readily identifiable, a lag phase could be discounted and, due to the nature of the process, growth would not be inhibited by surface drying or other factors.

Gill and Harrison (1985) extended the range of offal products examined to include heart, kidney, brain and sweetbread. Again, *E. coli* was selected as the marker organism. Its proliferation on offals was estimated by temperature function integration and determined directly from viable (plate) counts. After correction for the lethal effects of freezing on a portion of the population, the observed average values for *E. coli* growth were remarkably similar to the calculated values, and Gill and Harrison (1985) suggested that the hygienic efficiency of any offal cooling process could be quantified by the number of *E. coli* generations calculated from the product cooling curve. This view was reinforced by Gill (1986).

7.1.2. Meat Handling Processes

Smith (1985) examined the generation time, lag time and minimum temperature for growth of coliform organisms on meat with a view to understanding the rationale behind mandatory codes of practice used in commercial abattoirs.

Generation times and lag times of coliform growth aerobically on raw, blended mutton were estimated experimentally and found to agree closely with those calculated from the square root relationship. For generation time this was written in the form:

$$T = 3.40 + 18.58\sqrt{R} \qquad (7.1)$$

$$(r^2 = 0.98)$$

where T = temperature

R = generations h^{-1}

and for lag time:

$$T = 3.00 + 29.09\sqrt{(1/L)} \qquad (7.2)$$

$(r^2 = 0.96)$

where T = temperature

L = lag time (h)

These times were also found to agree closely with the times determined using a strain of *E. coli*. A strain of *Salmonella typhimurium* inoculated in the same way into blended mutton gave slightly longer generation and lag times at temperatures below 15°C. This indicated that the calculated values for lag and generation times could be used to determine the length of time raw chilled meat may be held at temperatures above the minimum temperature of growth without an increase in the number of any *Salmonella* organisms present. On this basis, Smith (1985) concluded that maintaining the temperature of boning rooms at 10°C or less was unnecessary providing the meat was processed within calculated time limits. At 10°C, 25 h would elapse before any increase in coliform numbers; at 15°C, 8.4 h and at 20°C, 4.1 h.

An important aspect of this study was the calculation of lag time variation with temperature. This was shown to be adequately described by the square root model with a T_{min} value similar to that estimated for generation time results (Fig. 7.2). Consequently, despite different actual rates at which lag time and generation time change in response to temperature, both can be described by the same **relative rate function** (see 6.2.) which depends only on the T_{min} value of the parameter studied (McMeekin *et al.*, 1988).

Equations 7.1 and 7.2 derived by Smith (1985) for coliform lag and generation times were tested by Smith (1987) by comparing results obtained by plate counts for *E. coli*, *S. typhimurium* and coliform organisms naturally present on sheep carcasses processed in a commercial abattoir.

Using three fluctuating temperature regimes (Fig. 7.3), the average temperature of mutton tissue was determined at 15 min intervals and the expected increases calculated using Eqn. 7.2 until this phase of growth was complete, and then the generation time equation was used. The average results obtained by Smith (1987) are shown in Table 7.1. Statistical analysis (*t* tests) indicated no significant differences between the results obtained by calculation and those obtained by plate counts.

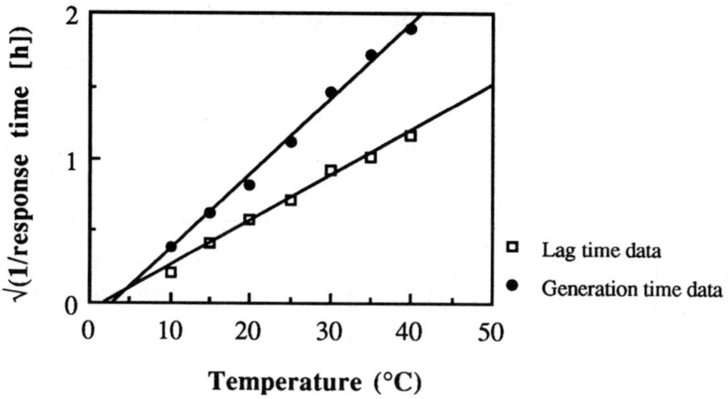

Fig. 7.2 Square root plot of growth rate and reciprocal of lag time *versus* temperature for *E. coli*. [Data of Smith (1985)]. Growth rate: Correlation coefficient = 0.994, T_{min} = 3.24 °C; Lag time: Correlation coefficient = 0.993, T_{min} = 2.08 °C.

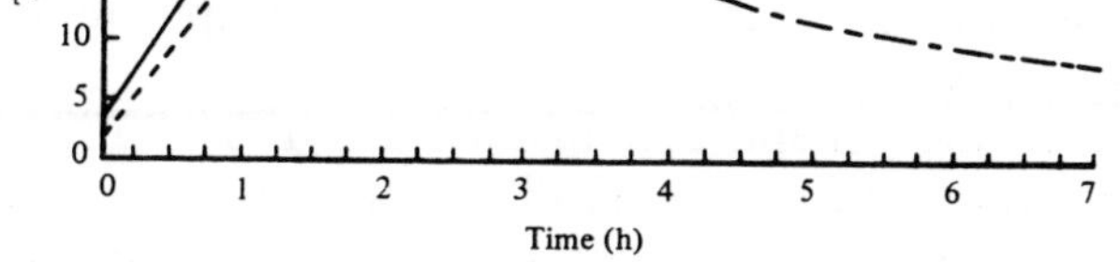

Fig. 7.3 Time-temperature regimes for the three water baths in which sachets of blended meat were incubated. [Reproduced from Smith (1987), with permission of Cambridge University Press].

Table 7.1 **Average Increase in Bacterial Numbers (Log_{10}/gram) in Sachets of Blended Mutton Tissue**[†]

	Calculated average increase[⊗]	Increase found in Coliforms (Mac.)	*E. coli* (Mac.)	*E. coli* (TYSG)	*S. typhimurium* (Mac.)	*S. typhimurium* (TYSG)
Bath A	1.11	1.08	1.09	1.08	1.02	0.96
Bath B	1.23	1.02	1.12	1.14	1.23	1.19
Bath C	1.00	0.97	1.07	0.95	1.07	1.08

† Temperature profiles for Baths A, B, and C are shown in Fig. 7.3, above.

⊗ Modified from Smith (1987).

Recovery of *E. coli* on a selective medium (MacConkey Agar) and non-selective medium (TYSG) was similar, indicating that coliform counts obtained in mutton tissue taken from carcasses in a commercial abattoir would provide a good estimate of the actual number of coliform bacteria present.

Furthermore, as the increases in numbers of naturally contaminating coliforms were similar to those obtained with pure cultures of *E. coli* or *S. typhimurium*, enumeration of coliforms would provide a reasonable estimate of potential increases of specific mesophilic indicator or pathogenic organisms.

Confirmation of the general conclusions drawn by Smith (1987) was provided by Mackey and Kerridge (1988). These workers considered the growth of *Salmonella* in minced beef, examined lag time and generation time, and considered the effect of inoculum size. The last criterion was included as the organisms of concern (*Salmonella*) in this situation are usually present in low numbers initially and must grow in competition with the normal spoilage microbiota of the food. The growth of *Salmonella* and other pathogens therefore may be influenced by the nature of the competing microbiota which may change with temperature, and by the initial ratio of pathogens to competing organisms.

Both high and low inocula obeyed the square root model and the combined generation time data was described by the equation:

$$\sqrt{R} = 0.0473T - 0.158 \qquad (7.3)$$

$$(r^2 = .986;\ T_{min} = 3.3°C)$$

where R = growth rate (generations h^{-1})
T = temperature (°C)

Similarly, the square root of the reciprocal of lag time was related linearly to incubation temperature, confirming the report of Smith (1985). Again, no significant differences were evident between results from the high and low inoculum experiments and the combined results were described by the equation:

$$\sqrt{(1/L)} = 0.0243T - 0.0711 \qquad (7.4)$$

$$(r^2 = .877;\ T_{min} = 2.9°C)$$

where L = lag time (hours)
T = temperature (°C)

Mackey and Kerridge (1988) noted that there was much scatter of lag time data about the fitted line. This may have been due to difficulties inherent in the accurate estimation of the completion of the lag phase, or to a differential rate of recovery to a physiologically active state between cells within a population, or between batches of cells used in different experiments.

Confirmation of the lack of effect of inoculum size, foreshadowed by Jason (1983), is an important result for the practical application of temperature function integration to describe the growth of pathogens on meat. The result is in keeping with that reported by Gill and Newton (1980) who observed no inhibition of *S. typhimurium* by a selection of spoilage organisms on meat and Gill and Newton (1977) who reported lack of microbial interactions on meat stored aerobically until maximum cell densities were reached. The experienced predictive microbiologists, R.L. Buchanan and T.A. Roberts (*personal communications*) also report that, in many foods, pathogens develop at rates determined by the formulation of the product and the conditions of storage. In most instances they appear to 'ignore' the normal spoilage microbiota. Under some conditions one can envisage situations where microbial interactions may occur, especially with *Lactobacillus* strains which produce bacteriocins or other antimicrobial compounds (Gill and Newton, 1980;

Collins-Thompson *et al.,* 1983). If these affect the growth rate of the organism of concern, this additional factor would have to be taken into account in developing predictive models. Once again, this emphasises the need to understand the microbial ecology of a product.

In addition to their own data, Mackey and Kerridge (1988) reviewed the effect of temperature on the growth rate of *E. coli, Salmonella* and coliforms (Table 7.2).

Growth rates in minced beef (Mackey and Kerridge, 1988) were significantly slower than in minced mutton (Smith, 1985), although the latter author reported similar results on the two substrates. The differences reported were probably due to differences in oxygen tension established in different experiments.

Despite differences in actual rates, similar T_{min} values were obtained for each of the data sets shown in Table 7.2 when analysed by the square root model. The mean T_{min} value (°C) was 3.43 ± 0.41 (SD). This suggests that the growth of *E. coli*, coliforms and *Salmonella* can be described by the same relative rate function. For each product it is necessary to determine actual rates, but a similar response to temperature fluctuations will occur. Table 7.3 shows relative rates at different temperatures based on a T_{min} of 277 K, if 10°C is taken to be the reference temperature corresponding to a relative rate of 1.00.

The data of Smith (1985) were selected by Ratkowsky *et al.* (1991) for a comparison of predictive models because the estimates of growth rate and lag phase duration were well replicated with between 4 and 14 determinations at each temperature. Replication is desirable if the best estimates of rate are to be obtained but unfortunately the constraints of time and labour imposed by many enumeration methods mean that replicated experiments in predictive microbiology are rare.

7.1.3. Meat Thawing Procedures

A further application of temperature function integration was considered by Lowry *et al.* (1989) in an attempt to determine the hygienic efficiency of meat thawing processes. To develop an appropriate model for the temperature dependence of growth rate and lag phase duration, *E. coli* was examined after exposure to freezing and thawing in simulated meat medium (SMM). This preliminary step, designed to approximate environmental insults occurring in practice resulted in reduction of *E. coli* numbers by 37-75 % indicating that 'wild type' gram negative pathogens in meat might be expected to survive freezing in significant numbers. After freezing and thawing in SMM, lag times and growth rates were determined at different temperatures. Validation

Table 7.2 **Mean Generation Times (h) of Salmonellae, *Escherichia coli* and Coliforms at Different Temperatures**[a]

		Temperature (°C)							
Organism	Growth Medium	10	15	20	25	30	35	T_{min}[b]	Ref.[c]
Cocktail of *S. thompson* and:									
	minced beef								
S. stanley	" (high inoculum)	15.2	3.03	1.59	1.11	0.65	0.47	3.69	1
S. infantis	" (low inoculum)	10.0	3.03	1.43	0.97	0.56	0.43	3.28	1
S. typhimurium	minced beef	9.65	2.65	1.60	0.75	0.49	0.39	3.78	2
Coliforms	minced beef	6.68	2.63	1.49	0.80	0.47	0.34	3.62	2
E. coli	minced beef	6.90	2.60	1.40	0.78	0.52	0.37	2.79	2

[a] Modified from Mackey and Kerridge (1988).

[b] Calculated from Eqn. 3.10.

[c] Reference 1 : Mackey and Kerridge (1988); Reference 2: Smith (1985).

Table 7.3 **Relative Growth Rates for a Mesophilic Organism, T_{min} = 4°C, T_{ref} = 10°C**

Temperature (°C)	Relative Rate
8	0.4
10	1.0
12	1.8
15	3.4
20	7.1
25	12.3
30	18.8
35	26.7

studies were carried out using inoculated meat slices to give the 'best fit' models.

A quadratic equation was used to describe the lag phase and a biphasic square root plot was selected to model growth rate with the 'break' occurring at 25°C. One possible explanation for the 'break' is that above 25°C growth was sufficiently rapid to cause anaerobiosis and a switch to fermentative metabolism, resulting in decreased growth rates. Whatever the reason, if the break occurs in practice it must be taken into consideration in the development of a predictive model. The goodness of fit of Lowry's three part model in carton thawing studies will be discussed in Section 7.3.

7.2. EFFECT OF VARIABILITY ON PREDICTING SAFETY (OR SHELF LIFE) OF A FOOD PRODUCT

Section 7.1. considered the development and validation of models to predict food safety. From the material presented in 4.4., however, it is clear that the ability to predict the generation or lag time of an organism or the shelf life of a product (as discussed in Chapter 6), depends strongly on the 'mean/variance' relationship of the probability distribution underlying these time-based variates. Consequently, presenting generation times with standard deviation values, as in Lowry *et al.* (1989), will be inappropriate unless the estimates are normally distributed. If the distribution follows an inverse Gaussian distribution, its scale parameter is $c = 4V(\sqrt{k})$ [see Eqn. 4.11], whereas if it obeys a gamma distribution, its scale parameter is $c = V(\ln k)$ [see Eqn. 4.12]. Consider the data of Neumeyer (1992) on generation time for *S. aureus* 3b, which appear in Appendix 4A.4. It was shown in 4.4.1. (see Fig. 4.6) that those data were better fitted by an inverse Gaussian distribution than by a gamma distribution. Therefore, one can fit these data using generalised linear regression with inverse Gaussian error, or one can fit the square-root model, Eqn. 3.10, in the 'usual' way by regressing $\sqrt{1/\theta}$ *versus* T, as described in 4.4. The latter approach leads to:

$$\begin{aligned} 4\,V(\sqrt{k}) &= 4\,(0.0002042) \\ &= 0.000817 \end{aligned}$$

which is close to the mean deviance of 0.00088 obtained (see Table 4.9) using generalised linear regression.

Using 0.000817 as the estimate of the scale parameter (although the mean deviance 0.00088 is an alternative choice), we have tabulated the mean predicted generation time $\hat{\theta}$ and associated predicted variance:

$$V = 0.000817\,(\hat{\theta})^3$$

as functions of temperature (Table 7.4). The $\hat{\theta}$'s were obtained from the predictions produced by the generalised linear regression model (see Table 4.10) but could also have been estimated using the 'usual' method, by adjusting those predictions for bias using the correction formula given for the inverse Gaussian distribution in Table 4.12.

The shape of the distribution of θ at 12.5°C , where the average growth rate, $k = 1/\theta$, is low, is highly asymmetric, there being a very long right-hand tail. On the other hand, at temperatures of 30-35°C, where the growth rate is high, the distribution of θ is much more symmetrical, and much closer to that of a normal distribution. Fig. 7.4 shows that the shapes of the distribution of θ at temperatures of 12.5°C and at 30°C, using the means and variances given in Table 7.5, which illustrates the differences in symmetry between the two temperatures.

Table 7.4 **Predicted Generation Time, $\hat{\theta}$ (min), and its Variance, V. [Data of Neumeyer (1992)].**

Temperature (°C)	$\hat{\theta}$	$V = 0.000817\,(\hat{\theta})^3$
12.5	887.76	571600.
15	388.62	47950.
20	138.23	2158.
27	56.43	146.8
28	51.03	108.6
30	42.32	61.9
35	28.28	18.5

It is important to keep the asymmetry of θ in mind when the object is to determine the safety (or shelf life) of a product. It would be quite misleading to produce a symmetric confidence interval of the form:

$$\hat{\theta} \pm t\ \mathrm{SE}\,(\hat{\theta})$$

where t is the critical value obtained from a table of the Student's t distribution and SE $(\hat{\theta})$ is the standard error of $\hat{\theta}$

such as one might do when $\hat{\theta}$ is normally distributed.

Instead, a different approach to the problem of estimating a confidence interval, or of putting bounds on the estimated generation time needs to be formulated. One way is to tabulate the times corresponding to pre-chosen proportions, equivalent to one-tailed P-values, of the cumulative distribution of θ. Thus, for distributions like the inverse Gaussian or gamma, if the parameters are known or can be estimated accurately from the data, one can find critical values of θ, denoted θ_0 corresponding to probabilities $P[\theta < \theta_0]$. For the data of Neumeyer

Table 7.5 **Values of θ (min) Corresponding to Certain Selected Probabilities $P[\theta < \theta_0]$, Inverse Gaussian Distribution with Scale Parameter c = 0.000817. Generation Time Data of Neumeyer (1992).**

12.5 °C		30 °C	
θ_0	$P[\theta < \theta_0]$	θ_0	$P[\theta < \theta_0]$
46	0.000001	17.75	0.000001
55	0.00001	19.25	0.00001
69	0.0001	21.0	0.0001
92	0.001	23.5	0.001
136	0.01	27.0	0.01
168	0.025	29.0	0.025
203	0.05	30.5	0.05
657	0.5	41.5	0.5
2356	0.95	56.25	0.95
2951	0.975	59.5	0.975
3782	0.99	63.75	0.99
6038	0.999	73.0	0.999

(1992) for generation times at 12.5°C and 30°C, the values of θ_0 obtained from the fitted inverse Gaussian distribution, using the statistical package GENSTAT (1989), are given in Table 7.5. At 12.5°C, one can interpret the result as indicating that if one wishes to ensure a risk of 10^{-3} or less a generation time of not more than 92 minutes should be used for predictive modelling. To achieve a more stringent risk of 10^{-6}, would require a more conservative maximum generation time of 46 minutes, and so on. At 30°C the rates are much faster and the generation times are much smaller, so that risks of 10^{-3} and 10^{-6} require generation times of 23.5 and 17.75 minutes, respectively.

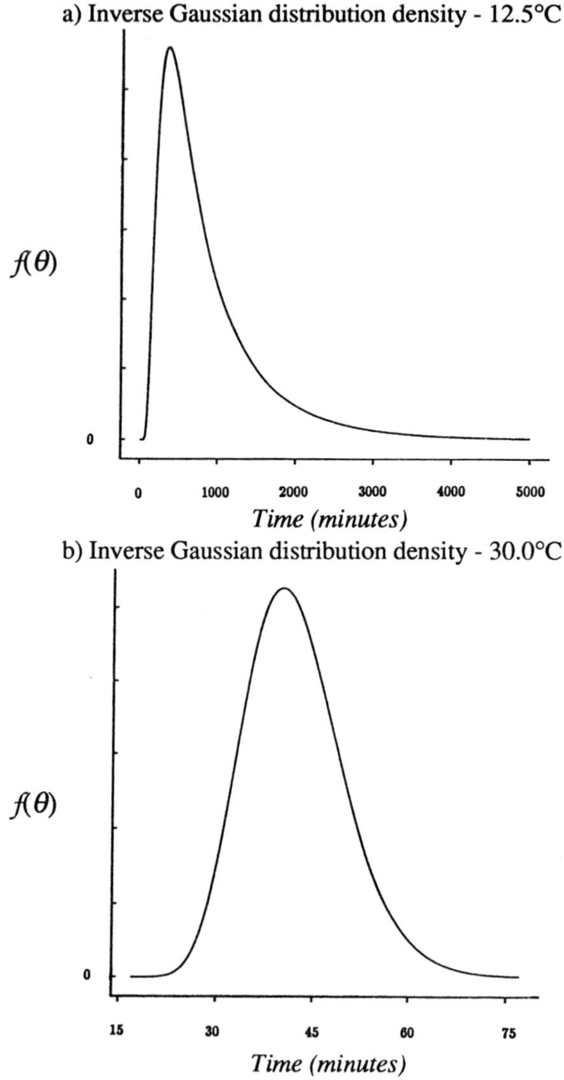

Fig. 7.4 Estimated probability density distribution function for inverse Gaussian distribution of generation time for θ at (a) 12.5°C and (b) 30°C for parameter values in Table 7.5, opposite. [Data of Neumeyer (1992), see Appendix 4A.4]

Similar exercises can be carried out with data that are gamma distributed, to produce bounds that attempt to take account of the variability of estimates of time-based quantities such as generation time and lag time. An interesting sidelight to the asymmetry of the distribution of θ under stringent growth conditions is the fact that some samples will have very long generation times. Thus, for the Neumeyer (1992) data, at 12.5°C one percent of the samples will have generation times of 3782 minutes (= 63 h) or more, and one in a thousand will require more than 100 h to double its population size.

7.3. PRACTICAL APPLICATION OF TEMPERATURE FUNCTION INTEGRATION TO EVALUATE PROCESS HYGIENE

The studies reported in 7.1. indicate the potential of temperature function integration to evaluate process hygiene. However, these dealt mainly with the development of appropriate models and their validation in meat products under well controlled laboratory conditions. Even under these well controlled conditions one must be aware of the effects of variability on predictions as discussed in 7.2., above. In addition, practical application must take into account variability in temperature history inherent in any processing operation. Gill *et al.* (1988b) listed difficulties in practical application as follows:

1) defining a process
2) collection and analysis of only those temperature data that are appropriate
3) ensuring that appropriate bacterial growth data are applied in each particular case, and
4) defining realistic criteria for permissible potential proliferation of *E. coli*.

An attempt to address these problems was made by Lowry *et al.* (1988) in their consideration of meat thawing processes. During thawing of cartons of boned-out meat, values for *E. coli* growth calculated from their model were usually within one generation of observed values at the top leading corner of the carton and from the centre top surface of the meat mass. At the latter sampling point, growth was always at least 10 generations less than that observed at the top corner. At the bottom leading corner observed growth averaged only 40% of calculated growth due to the rapid development of anaerobic conditions resulting from accumulation of drip from thawed surfaces.

These practical observations indicate the need to obtain detailed information on temperature changes and other variable factors to enable selection of an appropriate model and correct interpretation of the public health significance of the process.

C.O. Gill and colleagues reported the practical application of temperature function integration in a series of papers dealing with two types of carcass cooling procedures (Gill *et al.*, 1991a,b) and a hot boning process (Reichel *et al.*, 1991). For each type of process the experimental procedure involved collection of temperature histories from the slowest cooling, contaminated area of a random sample of units passing through the process. Temperature histories were integrated using models developed for *E. coli* to estimate the potential proliferation of mesophiles during a particular process and to characterise its hygienic adequacy in relation to conditions representing good manufacturing practice. For carcasses cooled by a traditional air cooling process the slowest cooling surface was in the aitch bone pocket, a region identified as important long ago by Scott and Vickery (1939), but any effect of surface drying also noted by these authors was ignored. Based on 50 cooling curves considerable variation in rates of temperature decrease were observed which had the potential to allow between 2 and 14 generations of *E. coli* with 80% of samples < 10 generations.

Because specification of a maximum permissible limit for proliferation was deemed inadequate to characterise an acceptable process, criteria for a three class attributes sampling scheme were suggested with the number of temperature determinations (*n*) > 20, the maximum permissible number of generations (M) = 14, the marginally acceptable limit (m) = 9 generations, and the number of samples in the m - M region (c) = 20%.

Application of the criteria proposed for the traditional side cooling process was shown to be applicable also to cooling of beef sides by spray chilling (Gill *et al.*, 1991b), and to cooling of cartons of hot boned beef in a blast freezer (Reichel *et al.*, 1991). The latter study provided an excellent example of the use of temperature function integration to identify defects in a process and to validate that upgrading the facility had produced the anticipated improvement in process hygiene.

On the basis of 50 temperature histories for the warmest microbially contaminated regions of product passing through both carcass cutting and carton cooling processes, the average proliferation of *E. coli* was estimated to be greater than the criteria values shown above for a traditional beef side cooling process. After upgrading a further 50 temperature histories showed that the average proliferation during the process had been reduced, and met the criteria for side cooling. The reliability of the model to predict proliferation was demonstrated by comparing calculated and observed values of *E. coli* growth which were within ±1 generation in 76% of comparisons.

7.4. SUMMARY

The studies of Gill and colleagues reported in this section demonstrate clearly the potential of temperature function integration to provide rapid, practical assessment of the hygienic adequacy of meat processing operations.

Furthermore, it is most instructive to have estimates of the degree of variability and distribution of cooling patterns encountered in practical meat processing operations. The collection and collation of such information provides an indication where to locate sensors to accumulate data of significance which can be interpreted in terms of the hygienic status of the product or process.

As with the questions raised by Smith (1987) on the validity of mandatory codes of practice requiring maintenance of boning room temperatures at <10°C, the work of Gill *et al.* (1991 a, b) opens to debate the rationale underlying public health regulations requiring cooling of deep muscle to 7°C before further operations.

8.

TIME-TEMPERATURE INDICATORS

Some of the commercially-available time-temperature indicators based on diffusion of waxes, enzyme-substrate kinetics and chemical reactions are considered. Their activation energies are compared with the changing 'apparent' activation energies of microbial growth.

8.1. DIFFERENT APPROACHES TO THE APPLICATION OF TIME-TEMPERATURE MONITORS

Theodore Labuza and his colleagues at the University of Minnesota have published extensively on the use of time-temperature devices to monitor food spoilage (Taoukis and Labuza, 1989 a, b; Fu *et al.*, 1991; Labuza and Taoukis, 1991; Sherlock *et al.*, 1991; Taoukis *et al.*, 1991). The 1991 papers appeared after preparation of material for this chapter and further illustrate the approach of the Minnesota group who have generally described and tested patented devices for their ability to simulate the kinetics of food spoilage.

The present authors approach the selection of monitors from a different philosophical perspective and prefer to first determine the temperature dependence of a deteriorative process. From that basis suitable monitors can be chosen for product descriptions. The latter approach is described in this chapter and has proved particularly useful in education and extension. Readers requiring greater detail on the operation of the various monitors are referred to the publications of Labuza and his colleagues cited above.

8.2. FULL HISTORY AND PARTIAL HISTORY INDICATORS

Wells and Singh (1988a) distinguished between "full history" indicators which respond continuously to time for all temperature conditions and "partial history" indicators which respond only for the period of time during which a temperature threshold has been exceeded. The basic principle of a 'full history' time-temperature monitor is that its reaction responds to the hypothesis of time-temperature tolerance described in 6.3. Taoukis *et al.* (1991) have called partial history monitors "Critical Temperature Time Integrators" when they reflect the cumulative time-temperature above a reference critical temperature.

8.3. SELECTION OF FULL HISTORY MONITORS

Most of the monitors used to simulate bacterial growth or food spoilage are based on Arrhenius kinetics while, as preceding chapters have shown, the Bělehrádek temperature function is more appropriate. Fu *et al.* (1991) accepted the premise that the activation energy of microbial growth could be changing continuously but felt that this would be difficult to model. In order to envisage the suitable activation energy of a monitor over the desired temperature interval, it may be useful to review the differential equations, discussed in 3.4, which McMeekin *et al.* (1988) used to reconcile the Arrhenius law and the square root model. Apparent activation energies appropriate to bacterial growth are presented in Table 8.1, which is reproduced from McMeekin *et al.* (1988) with the term E_{app} substituted for E, the activation energy.

At temperatures close to T_{opt}, i.e. at temperatures where $T - T_{min}$ in Table 8.1 are approximately 30°C, E_{app} approximates to E, the 'true' activation energy of the rate limiting enzyme for growth and metabolism. As the temperature is lowered decreasing values of $T - T_{min}$ produce greatly increased apparent activation energies for a 5°C temperature interval. The activation energy of the native enzyme is deemed to remain fairly constant while the apparent activation energy increases markedly due to the denaturation of much of the enzyme. It may be more appropriate to consider the change in apparent activation enthalpy ΔH_{app}, which is related to ΔE_{app} by the equation:

$$\Delta E_{app} = \Delta H_{app} - RT$$

The effect of the 'RT' term is so small, however, that for economy of space a second table is not justified. It is clear from Table 8.1 that ΔH_{app} would change continuously as predicted by Brandts (1967). Table 8.1 may be considered as a 'rule of thumb', and readers who favour other models to describe the temperature dependence of bacterial growth may prefer to use these to derive values for E_{app} upon which to determine the usefulness of different time-temperature monitors for their purposes.

Relative rates of spoilage are easier to envisage than apparent activation energies. In subsequent tables the data will be represented as relative rates (described in 6.2.). Relative rates calculated by the Bĕlehrádek power function (Eqn. 3.10) for psychrotrophic spoilage were shown in Table 6.2, and for mesophilic spoilage in Table 7.3. The reconciliation equation (Eqn. 3.24), however, tells us that at each relative rate temperature interval we are looking for a monitor with a similar activation energy for that interval.

Table 8.1. **Apparent Activation Energy (E_{app}) Expressed as a Function of T-T_{min} for Four Selected Bacteria**

Equation 3.24:

$$E_{app} = 2RT^2/(T - T_{min}),\ \text{kJ mol}^{-1}$$

$T - T_{min}$	Psychrophile $T_{min} = 251$	Psychrotroph $T_{min} = 263$	Mesophile $T_{min} = 278$	Thermophile $T_{min} = 308$
5	218	239	266	326
10	113	124	138	168
15	78.4	85.8	95.0	115
20	61.1	66.5	73.6	89.5
25	50.7	55.2	61.1	73.6
30	43.8	47.7	52.7	63.2

8.3.1. Full History Monitors Based on Enzymes

I-POINT® AB of Sweden pioneered the manufacture of a biochemical time-temperature indicator which was developed by Kockums Chemical AB, Malmo, Sweden (Blixt and Tiru, 1977). Unfortunately, I-POINT® AB have recently ceased production and organisations such as the World Health Organisation (WHO) are seeking other companies which could produce the same type of product. It is therefore worthwhile to discuss the concepts in some detail. Any company wishing to produce such monitors would be wise to target specific markets. An I-POINT® monitor is shown in Fig. 8.1.

Before activation the device contains a two-part, inner ampoule, one part containing an enzyme solution and the other a substrate solution. Another key ingredient is a liquid pH indicator. The reaction is started by applying pressure on the package, whereupon the barrier between the two parts is broken and their contents mixed. The self-adhesive indicator strip is then attached to the product to be monitored. The change in the speed of the reaction is effected by the use of different enzymes and substrates, while the parallel displacement of all TTT (see 6.3) curves is manipulated by the use of buffers. The enormous flexibility of I-POINT® indicators is shown in Fig. 8.2.

Most workers have selected I-POINT® monitors from the manufacturer's product catalogue and tested them against spoiling products (Mistry and Kosikowski, 1983; Wells and Singh, 1988a). As stated in 8.1, the present authors have preferred to work on the theory of spoilage models and consider the types of monitor appropriate to those models which are already available, or which should be purpose designed for the product.

Table 6.1 showed the relative rates of spoilage of proteinaceous foods at chill temperatures based on the Bělehrádek temperature function (Eqn. 3.10). The table has only been calculated up to 15°C because above this temperature mesophilic organisms begin to affect the relative rates.

Table 7.3 showed the relative rates of growth of mesophilic enteric organisms such as *E. coli* and Salmonellae, also calculated by the use of Eqns. 6.1 and 3.10.

Let us now take pasteurised milk as an example for selecting a monitor. The use-by date of Australian milk is typically taken as 8 days at 4°C. Therefore we are looking for a type of indicator which would change to colour 2 at 4°C in 8 days. The two indicators in the I-POINT® AB Product Manual which change to colour 2 (100% of TTT) after 8 days at 4°C are models 3270 and 4007. Taoukis and Labuza (1989a) selected type 4 models for study because they favour Arrhenius kinetics. Fig. 8.3 shows that I-POINT® Type 4 strips show excellent Arrhenius

Interpretation A. For this model of I-POINT®TTM the label has a pre-printed color scale to be used as reference for four defined colors appearing in the window during the active life of the TTM

and as can be seen on these labels, the meaning of the colors can be expressed either verbally or numerically.

Our presently recommended interpretation of the color signals are:

As long as Color 0-EXCELLENT or 1-GOOD appears;

80 % or less of the preset TTT is used up.

When Color 2-UNCERTAIN appears;

100 % of the preset TTT is used up.

When Color 3-OVEREXPOSED appears;

130 % or more of the preset TTT is used up.

Fig. 8.1 Reproduction of TTT monitors (actual size) from I-POINT® Product Manual. Colour 0 is green; colour 1, fawn; colour 2, brown; colour 3, orange. The criteria for use of TTT monitors are described in 6.3.

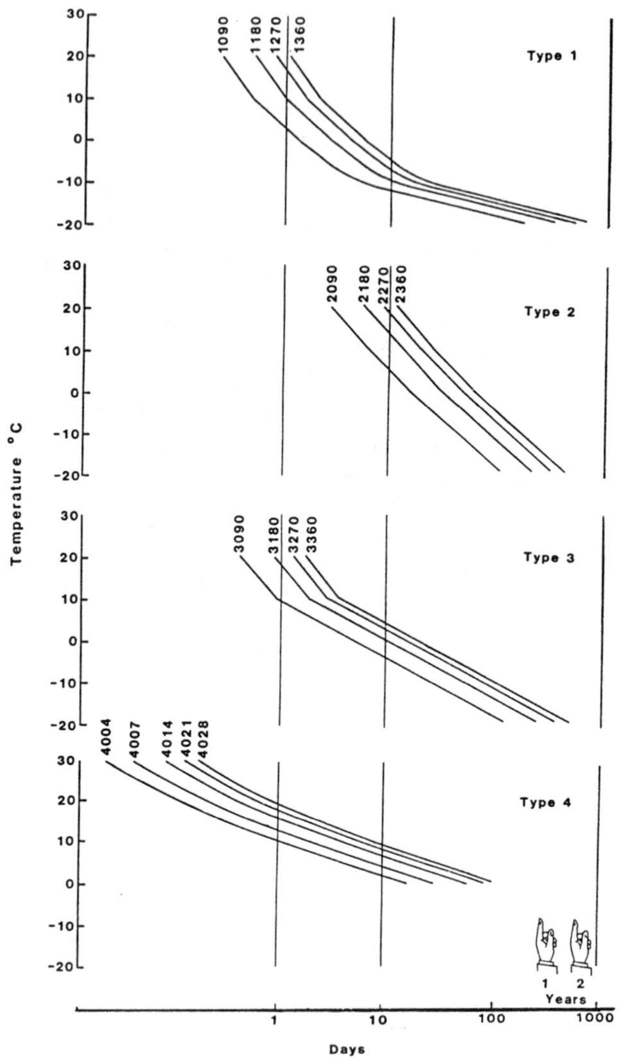

Fig. 8.2 Time-to-colour change 2 for four types of I-POINT® TTM strips. [Reproduced from Bremner *et al.* (1987) by permission of Elsevier Science Publishers BV, Amsterdam].

kinetics. The calculated activation energies of ~139-144 kJ mol^{-1} are in agreement with the experimental values obtained by Taoukis and Labuza (1989) for model 4021. Table 8.1 shows that Type 4 strips would be suitable for monitoring mesophiles approximately 10°C above T_{min}. The present authors, however, would not consider this model suitable for simulating spoilage of proteinaceous foods at chill temperatures. The Arrhenius plot of the Type 3 indicators is shown in Fig. 8.4, overleaf, in which the characteristic concave downwards plot of a Bělehrádek power function appears. A square root plot of the data from the Manual (Fig. 8.5) shows that the Type 3 models obey Eqn. 3.10 with a T_{min} of -8.8°C, very close to the observed value of -8.5°C for pasteurised milk. The characteristics of the Type 3 Model in the I-POINT® Manual show that the enzyme is active until at least -20°C. However, Fig. 8.4 and Table 8.2 show that this is the indicator of choice.

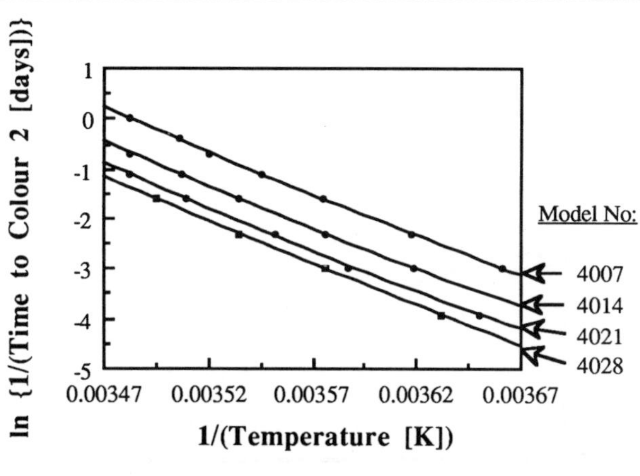

Fig. 8.3 Arrhenius plots for Model A Type 4 I-POINT® indicator strips at chill temperatures (0 - 15°C). Data read from the I-POINT® Product Manual. (Data: • ; Fourth degree polynomials fitted by CA-Cricket Graph[1] to data in the temperature range ~ 0 - 30°C: ——).

[1] Computer Associates International, Inc., 1240 McKay Drive, San Jose, CA, USA.

It will be seen that indicator 3270 more closely simulates milk spoilage than 4007 at the higher temperatures. McMeekin and Olley (1986) had already selected I-POINT® 3270 for monitoring of fish spoilage as 14 days at 0°C is very appropriate for the limits of edibility of temperate water species (Lima dos Santos, 1981). The T_{min} is approximately 1°C higher than that quoted for fish, meat and poultry in Table 6.1, and would have the effect of overestimating the relative spoilage rate between 0°C and other temperatures by approximately 14%. Ironically, Sherlock *et al.* (1991) rejected the Model 3 type I-POINT® indicators because they did not obey Arrhenius kinetics, the very reason that they were chosen by the Tasmanian group. The I-POINT® company sold indicator strips in rolls of 1000. These were transhipped refrigerated and had a use-by date themselves, i.e. unactivated TTMs had a limited time/temperature shelf life. Re-testing of indicator 3270 purchased by the CSIRO Seafood Technology Section a week before the use-by date indicated a small change in response, but this was attributed to changes in lighting. Wells and Singh (1988b) have already commented on the

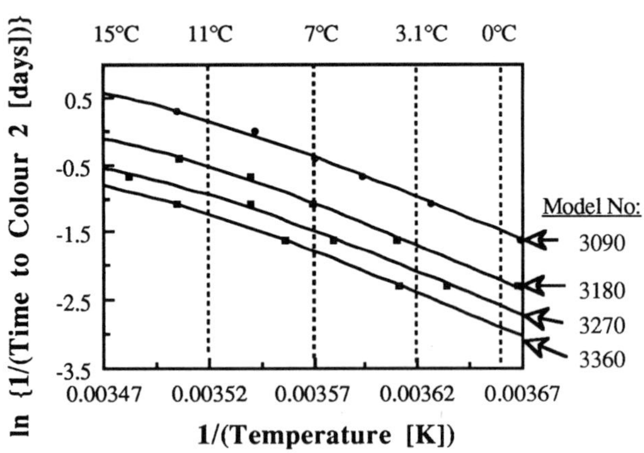

Fig. 8.4 Arrhenius plots for Model A Type 3 I-POINT® indicator strips at chill temperatures (0 - 15°C). Data read from the I-POINT® Product Manual showing concave downwards curves typical of a power function. (Data: • ; Fourth degree polynomials fitted by CA-Cricket Graph[1] to data in the temperature range ≈ 0 - 30°C:—).

Table 8.2. **Performance of I-POINT® Indicators Compared with the Rate of Spoilage of Pasteurised Milk Using the Rate at 0°C as a Standard and a Use-By Date of 8 Days at 4°C.**

Temperature (°C)	Relative rate of milk spoilage	Time to use-by date (days)	Time(days) to 100% TTT of indicator 3270	4007
0	1	17.1	14	20
4	2.14	8	8	8
8	3.68	4.6	3.75	3.6
12	5.67	3.0	2.5	2.0
16	8.09	2.0	1.75	0.6

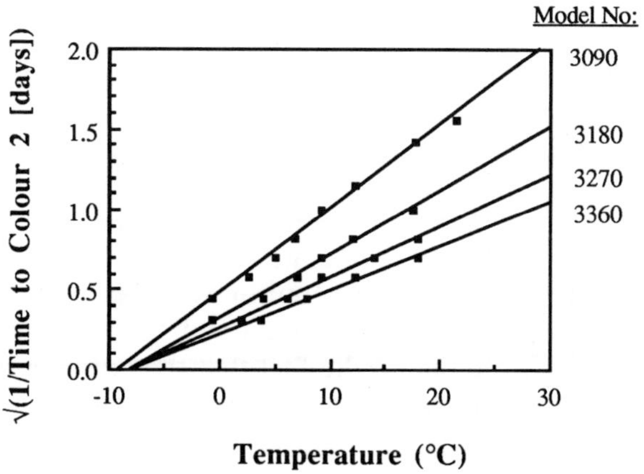

Fig. 8.5 Bělehrádek plot for model A Type 3 I-POINT® indicator strips showing that Type 3 strips obey equation 3.10. Data read from the I-POINT® Product Manual.

discrepancies caused by the visual perception of the observer and recommended the development of an electronic colour detector and the definition of the I-POINT® response scale in terms of the detector signal. Ultimately I-POINT® strips stored for too lengthy a period appeared to react at a faster rate due to increased initial product levels as the ester spontaneously hydrolysed to titrateable acidic products without mixing with the enzyme.

A psychological advantage of the I-POINT® indicators was that they simulated microbial growth: a 'lag phase' while the indicator colour did not change very rapidly followed by a rapid change in colour near the end-point of the indicator (exponential phase). As will be seen later, monitors based on diffusion cannot have this property.

Grisius *et al.* (1987) comment on the usefulness of the lag caused by the lack of a discernible colour change until sufficient substrate has been hydrolysed. They noted the lag in development of psychrotrophs in pasteurised milk and obtained significant correlations between I-POINT® indicator responses and psychrotrophic counts at 0°C and 5°C, but not at 10°C. They were, however, using Type 2 monitors which we would not have recommended.

8.3.2. Full History Monitors Based on Chemical Reactions

Wells and Singh (1988b) have reviewed their experience with a chemical monitor based on the polymerization of an acetylenic molecule. The Life Lines Freshness Monitor is produced by Lifelines Technologies, Morristown, NJ 07960 (Fields and Prusik, 1986). The indicator response is monitored with a hand-held microcomputer and optical light wand which records the reflectance (0 to 100%) of the polymer.

Table 8.3 shows their calculated mean value of the reaction rate constant for model Life Lines 57 compared to the relative rates of spoilage of pasteurised milk and temperate water fish species.

It can be seen that Life Lines 57 Monitor is better for fish than milk up to 10°C, but more suited, though not good for pasteurised milk at 15°C. The discrepancies are due to the fact that the polymerization of the acetylenic compound is obeying Arrhenius kinetics and not Bĕlehrádek temperature functions. Taoukis and Labuza (1989a) tested models 41 and 68 and found activation energies of 85.8 kJ mol^{-1} and 82.4 kJ mol^{-1}, respectively. Table 8.1 shows that these activation energies are typical of psychrotrophs at temperatures around 25°C above T_{min} (15°C - (-10°C)) and do not simulate chill temperatures for food storage.

Table 8.3. **Polymerization Rates of Compounds in Life Lines 57 Monitor at Food Spoilage Temperatures Compared with Spoilage Rates for Pasteurised Milk and Fish.**

Temperature (°C)	Reaction rate constant (day^{-1})	Rates relative to rate at 0°C	Fish (T_{min}= - 10°C)	Milk (T_{min}= - 8.5°C)
0	0.01526	1.00	1.00	1.00
5	0.03427	2.25	2.25	2.49
10	0.05127	3.36	4.00	4.65
15	0.1290	8.45	6.25	7.47

The envisaged Life Lines Inventory Management System is shown in Fig. 8.6 overleaf. The hand held microcomputer is not cheap and the system is only envisaged for warehousing and transport.

8.4. PARTIAL HISTORY MONITORS BASED ON DIFFUSION

The 3M Monitor Mark® time-temperature monitors are based on the migration of coloured fluids through a porous medium (Manske, 1983). The migrating chemical is held in a reservoir pad separated from the porous wick by a tag. The migration rate of the dyed chemical along the wick is a function of the amount of chemical contained in the reservoir pad. Taoukis and Labuza (1989a) showed that the migration of coloured fluid is governed by the laws of diffusion, and the Arrhenius kinetics of diffusion therefore apply to the system. The migrating fluid is usually a fatty acid ester such as octyl octanoate with 0.5% of a blue dye incorporated. Taoukis and Labuza (1989a) used a fatty acid ester with a melting point of -9°C, the monitor having been purpose-made for the study. For chilled foods it would thus be a full history monitor. The Arrhenius activation energy of this model was found to be 41kJ mol^{-1}

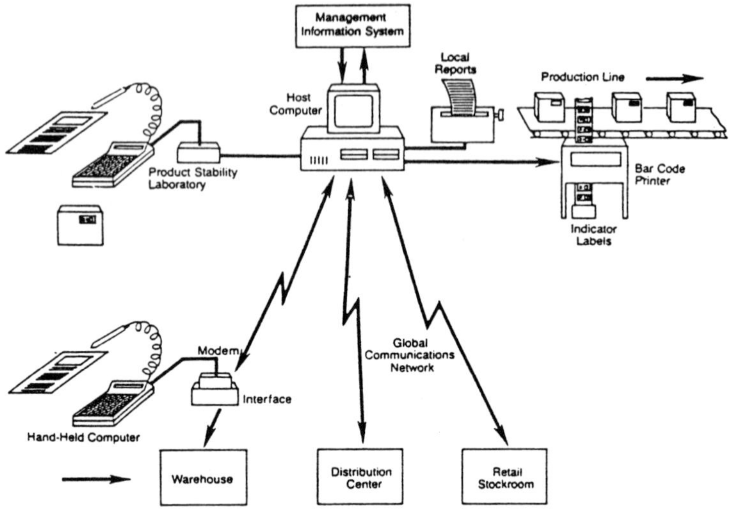

Fig. 8.6 A schematic diagram of the Life Lines Inventory Management System. Remote scanners are used to monitor product freshness in the distribution system. Freshness data can be downloaded to a host computer for inventory rotation decisions based on product quality. [Reproduced from Fields and Prusik (1988) by permission of Elsevier Science Publishers BV, Amsterdam].

between 0°C and 30°C. Table 8.1 shows that it would only be suitable for monitoring bacterial growth close to the optimum growth rate. These monitors cannot simulate bacterial growth at chill temperatures, only gross temperature abuse.

8.5. DO TIME-TEMPERATURE INDICATORS SIMULATE MICROBIAL SPOILAGE?

Taoukis and Labuza (1989a) have shown that the common activation energies encountered in food spoilage are covered by diffusion type enzymic and chemical time-temperature monitors. Their paper does not consider two very important factors:

(1) The enzymes concerned in bacterial growth are progressively denatured as temperatures approach T_{min} and thus a monitor is required to simulate activation energies of protein denaturation in the range of 200-300 kJ mol^{-1}.

(2) Spoilage products in foods are usually the result of bacterial metabolism, and are directly proportional to the amount of bacterial growth (see 3.3.2.). The activation energies required therefore vary from 40-300 kJ mol^{-1} (Table 8.1). The enzymic monitors are those with the greatest potential for manipulation, and though it is not clear how it has been achieved, Blixt (1983) shows that I-POINT® Type 3 monitors have a marked change in activation energy below and above 10°C. Fig. 8.3 has shown that this type can be considered to approximately obey a Bĕlehrádek power function between 0°C and 20°C.

Food Control AB (1974) patented a method of manufacturing devices with purpose-built characteristics. Quoting from the patent, Food Control AB state that the basis of the system is: "*the use of an enzymatic process in which an enzyme and a substrate of an enzymatic indicator react together, resulting in the formation as a function of time and temperature, of a reaction product that is utilised for indication. One of the substances employed for this purpose is urease which is allowed to act upon urea to form ammonia, the presence of which is established by means of colour change of a pH indicator, the temperature and the time logarithm being linearly interrelated. By adding a buffer, the time over which the pH indicator changes colour can be prolonged, i.e. the linear inter-relation is parallely off-set. However, the inter-relation obtained by means of such an enzymatic process is not always the desired one. For instance, if a suitable inter-relation exists between the time of the colour change at low temperatures, the time for the colour change at higher temperatures will frequently be too long*".

Expressing this another way, the enzymatic reaction at higher temperatures represents the part of the Arrhenius curve where the enzyme is performing close to T_{opt} while at lower temperatures the enzyme is partially reversibly denatured and is effectively present at a lower concentration. However, to clarify the concept in the patent, it is better to think in terms of the enzyme reacting too slowly at the higher temperature. The patent goes on to describe a second chamber in the device separated by a semi-permeable diaphragm (the permeability of which increases with an increase in temperature) or a bimetallic strip. In the second chamber is the product of the enzyme-substrate reaction, in this case ammonia. So in effect we have simulation of the optimum Arrhenius kinetics and enzyme denaturation by chill temperatures.

How much 'correction' of the urease system is required to simulate bacterial growth depends on a comparison of the Gibbs free energy of the enzyme urease and that of the key enzyme in bacterial growth (*cf.* Chapter 10).

8.5.1. Application of Partial History Indicators in the Food Industry

The 3M company has produced a wide range of partial history indicators. The reservoir pads contain a range of fatty acid esters which melt at a range of temperatures. As the esters do not progress along the wick when they are solid, only times of abuse are accumulated.

The company recommends a variety of indicators for the food industry which are shown in Table 8.4.

The tags are supplied in boxes of 100 with a card showing the time/temperature relationship of the tags supplied. A typical tag with reference card is shown in Fig. 8.7. These tags have been tested for the WHO by the CSIRO Seafood Technology Section, Hobart, and have been found to be extremely accurate. They in fact highlighted the uneven air flow even in forced draft incubators. The predicted response was obtained by carving a wooden torus with 50 grooves at equal intervals. The tags were placed in the grooves with the reservoir pads to the centre and thermocouples were placed at the centre of the ring.

While 3M Monitor Mark tags have been tested by research groups, there are few reports of trials in the market place. A cursory survey by the CSIRO Seafood Technology Section, Hobart, showed that local supermarkets had the capacity to keep packaged fish so that the blue dye would not even appear in window 1 after a week's storage. Local by-laws which prevent the storage of fish and meat together in one display cabinet, however, resulted in the minor product being displayed under inadequate refrigeration so that a 5I monitor in some supermarkets

Table 8.4 **Some Monitor Mark™ Time/Temperature Tags Recommended for the Food Industry By 3M**

Tag*	Response temperature (°C)	Application
5 L	5	Refrigerated products
5 I	5	Prepared foods
L 1	-1	Pre-plated meats

*L designates a maximum exposure of 7 days; I designates a maximum exposure of 48 hours. The letter before the number indicates a reference temperature below 0°C.

showed all 5 windows blue in 48 hours. These tags are thus an excellent teaching and management tool. The authors have found them useful for monitoring the amount of potential storage life of packaged chilled fish used up in the boots of motor cars, a weak link in the food chain from retailer to home refrigerator (James and Evans, 1990) and in assessing the claims for 'cold retention' of shopping bags made from materials (aluminium laminates) which are spin-offs of the space programme.

8.6. MONITORING VACCINE SHELF LIFE

3M Monitor cards have been exploited by the Expanded Programme on Immunization of the World Health Organisation to monitor the shipment and storage of vaccines. The food industry could learn much from their approach. A typical monitor card is reproduced in Fig. 8.8. It is interesting to note that polio vaccine is the most sensitive to temperature abuse. These multiple monitor cards have been based on accelerated stability tests using the Arrhenius equation.

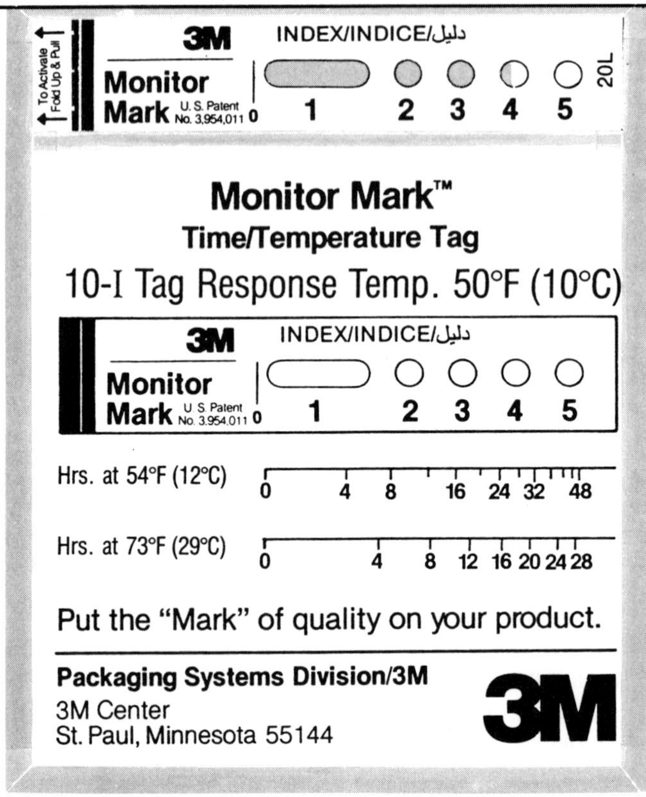

Fig. 8.7 3M Monitor Mark™ temperature history indicator strip (upper). A reference card for interpretation (below) is included with each box of 100 monitor strips.

The WHO had anticipated that the price of monitors would decrease with the increased scale of production. The theory is that, usually, the greater the turnover of a product the lower the price. Inflation has negated this hope and they are seeking cheaper alternatives. Cold Chain Newsletter (1989) notes that despite their consumption of 3M Monitor Marks rising from a few thousand in 1979 to over two hundred thousand in 1990, the price per tag has doubled. Cheaper alternatives to 3M do not perform as expected.

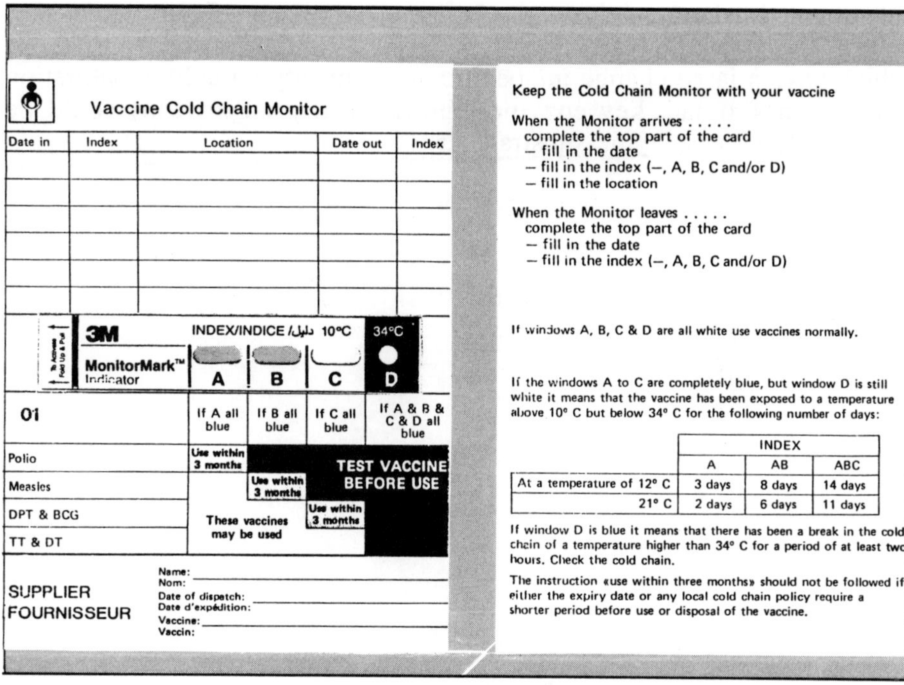

Fig. 8.8 Vaccine monitors used by WHO Expanded Programme on Immunization (1989). In this example windows A and B are completely blue. Thus: Polio vaccine should be tested before use; Measles vaccine should be used within three months; and DPT & BCG and TT & DT vaccines may be used safely. [Reproduced with permission of Expanded Programme on Immunization, WHO, Geneva].

Unless new, cheap monitors become available, it is wise to consider the new temperature loggers programmed to simulate the required spoilage mode which can be custom-built and are currently appearing on the market. These will be discussed in the next chapter.

8.7. SUMMARY

There are currently no cheap visual monitor strips available which simulate bacterial growth or food spoilage over a wide enough temperature range. One type, marketed until recently by the I-POINT®

company, showed promise and it is hoped that their technology can be rehabilitated. Alternatively, in the future, enzymes specifically selected to mimic the bacterial growth-rate-temperature relationship (e.g. those which have a large change in specific heat capacity on cold denaturation and which simulate key enzymes limiting bacterial growth) could be incorporated into enzyme-substrate strips.

9.

ELECTRONIC TEMPERATURE FUNCTION INTEGRATION

Basic concepts regarding mathematical integration, electrical transducers and electronic memory are presented to enable a discussion of the potential for electronic predictive microbiology devices. The development of temperature function integrators is also briefly traced and contemporary electronic predictive microbiology devices reviewed.

9.1. INTRODUCTION

The potential advantages of predictive food microbiology have been considered in earlier chapters. In particular, there is the promise of a rapid, non-invasive, and objective means of assessing the effect on product quality and safety of processing, storage and distribution conditions without recourse to traditional (culture based) microbiological skills and facilities. One of the most obvious applications of predictive microbiology is the prediction of elapsed or remaining shelf life of foods, although the approach is applicable to any perishable commodity (e.g. drugs, vaccines), because it is easier in almost all situations to measure environmental parameters and relate these to bacterial growth than to measure that growth directly. A single 'environmental history', moreover, can be analysed by many models specific for groups or species of interest to predict the extent of their growth. Using existing methodology, groups or species of microorganisms of interest must be isolated and/or enumerated separately.

Previous chapters have dealt with the development and validation of mathematical models for predictive microbiology. Buchanan (1991) considers that "while a majority of these models provide reasonable estimates of the potential for microbial growth in food systems, there has been relatively little transfer of this technology to food microbiologists involved in non-research aspects of the field" and further "that the limited acceptance of predictive modeling techniques seems to be due in large part to a lack of application software that reduces to routine operations the often complex mathematical manipulations associated with the use of the models".

The concept of temperature function integration for predicting bacterial growth was envisaged by Scott (1936, 1937), and was again considered in the 1970's by Nixon (1971) who presented, and subsequently patented, working designs for predictive devices. Little realisation of the technological potential, however, has yet been achieved. That which is currently available includes interactive computer software developed at the US Dept. of Agriculture Eastern Regional Research Centre in Philadelphia (Buchanan, 1991) which enables prediction of pathogen growth under constant storage conditions. Other technology which monitors and records the *fluctuating* temperatures experienced by foods in commercial practice, and interprets those temperature histories as the predicted growth of microorganisms of concern, is also available for a number of specific applications. Earlier reviews of this technology were published by Olley (1978) and McMeekin and Olley (1986).

These contemporary devices and the potential for application of electronic technology for predictive microbiology are the topic of this chapter. Readers well-versed in integration, electronics and computing will find the sections entitled 'Hardware Technology' and 'Integration' elementary. A starting point for this discussion is that microbial growth responses to environmental conditions can be modelled reliably, at least within defined parameters. The validity of this assumption was discussed in Chapters 6 and 7.

9.2. HARDWARE TECHNOLOGY

A workable predictive microbiology instrument must have a reliable method of measuring environmental factors affecting growth rate and, because we are interested in the cumulative effect of those factors on product quality and safety over time, it must also be able to convert the parameters measured into a form which can be efficiently stored and retrieved at some later time. Preferably the device should also be able to take those data and analyse them to predict the extent of microbial growth during that product's storage history. Overall then, a predictive microbiology device would be expected to incorporate suitable sensors, some means of converting the sensor output into a 'storable' form which can be mathematically manipulated, a method of storing the recordings and some facility for analysis and data interpretation. Furthermore, as we are concerned with *rates* of microbial growth, some form of timer is also essential.

Though other methods of measurement and data storage are available (for example, a material which significantly expands or contracts in response to temperature connected to a chart recorder will provide a continuous record of temperature), the advent of the integrated circuit (IC) and semiconductor technology have meant that for practical predictive microbiology devices the above mentioned elements are almost certain to be electrical transducers, analogue to digital converters, solid-state memory chips, a suitably programmed microprocessor chip to coordinate data storage, retrieval and analysis, and an electronic clock chip. The compactness, low power consumption, robustness, comparatively low cost and 'programmability' of semiconductor technology is such a powerful combination that it is unlikely that any new predictive device would use older technology.

9.2.1. Sensors

A variety of sensors exist which are able to measure environmental parameters relevant to microbial growth rate and present these as electrical signals. They are collectively termed **transducers**. Probably the most established transducer technology is in thermometry, although pH measurement is also well established and compact relative humidity, oxygen, and ion-selective probes have also been commercially available for some time. The following discussion is intended to serve as an introduction to some of the terminology likely to be encountered, and to impart an appreciation of the range of possibilities for monitoring the growth environment of microorganisms on foods. More comprehensive treatments may be found in Sydenham (1985) and Usher (1988) from which most of the following information is drawn.

9.2.1.1. Temperature

Four main types of temperature transducer are in contemporary use: **thermocouples**, **resistance thermometers**, **thermistors** and **silicon transistors**[1]. Thermocouples consist of an electrical circuit comprising two dissimilar metals which generate a current if the two metal junctions are exposed to different temperatures. The voltage generated is proportional to the temperature difference, thus a constant

[1] A transistor is a solid-state circuit component which can be used to switch or amplify a signal.

reference temperature or a device which simulates one, is necessary. Various metal combinations allow temperatures in the range ≈ -250 to ≈ +2800°C to be measured. Thermocouples can achieve sensitivities of 0.001°C, although those in routine use would seldom have sensitivity greater than 0.1°C. They are manufactured in a wide range of forms including very small devices. Platinum resistance thermometers are the most accurate of temperature transducers. They rely on the linear change of electrical resistance of a platinum wire element with temperature, are stable with time and require no reference or standard other than an accurate voltage source, usually a standard cell. A thermistor (thermal resistor) is a semiconductive material, the electrical resistance of which is very sensitive to temperature. The resistance is logarithmically related to temperature but this can be accommodated by appropriate circuit design. Thermistor thermometers are able to measure temperatures from the cryogenic range to over 300°C, using three different groups. They are available in a range of packaging from small beads to flat discs, and may be as small as 0.5 mm in diameter. Silicon transistor-based temperature transducers are discussed later.

9.2.1.2. Water activity

A range of devices is also available for the determination of relative humidity, i.e. water activity. The method of choice for $\mathbf{a_w}$ determination is the dew point method which involves monitoring changes in the reflectance of a mirror surface due to condensation as temperature is decreased. Other non-transducer methods exist also, and though most could be combined with other transducers to yield indirectly an electrical signal related to $\mathbf{a_w}$ the extra bulk and special conditions required make them unlikely candidates for operation in the field. A number of devices, called **hygristors**, are available in which resistance is logarithmically proportional to relative humidity/$\mathbf{a_w}$. Yet other transducers make use of capacitance changes with $\mathbf{a_w}$. Hygristors rely on the fact that the resistance of a moist substance connecting two electrodes is dependent upon air moisture content. Types include the evaporated grid sensor (gold fingers interleaved over an inert substrate), the Warren sensor (a humidity sensitive conducting plastic grid with a special plastic backing) and the Dunmore sensor (two electrodes surrounding a hygroscopic salt in equilibrium with an ambient atmosphere). A capacitance based $\mathbf{a_w}$ transducer for use in air has been made using porous anodised aluminium sandwiched between layers of gold which are the electrodes. This acts as an aluminium oxide capacitor in which moisture diffuses into the pores.

Problems with $\mathbf{a_w}$ transducers include temperature and hysteresis effects; the range of $\mathbf{a_w}$ over which the devices are accurate; and the sensitivity which is usually only ± 1 or 2 per cent. The Dunmore sensor is currently the most widely used in food microbiology, and has been employed in a variety of configurations which have permitted multiple point remote sensing, in-line measurement of free flowing materials and even sensors small enough to be mounted within a hypodermic syringe. (Troller and Christian, 1978). Lithium chloride is the salt normally employed although anodised aluminium (Al_2O_3) has also been used. Their disadvantages include inaccuracy, e.g. due to contamination by polar compounds, hysteresis at high humidities, aging effects due to physical and chemical changes, water droplet splashes on the sensor, as well as high cost and the need for careful calibration. Some of these problems can be overcome by good design and the use of absorbent filters and mechanical barriers to protect the sensor. Capacitance devices have the advantage of relative insensitivity to temperature in the range 0-80°C and negligible hysteresis. Hygristors have found widespread application in industry and laboratories due to their relatively good precision, ease of operation and reasonably rapid determinations.

9.2.1.3. Electrochemical sensors

A variety of chemical species involved in mediating shelf life can be quantified using transducers. Two plates suspended in a liquid form a primary cell, and a voltage occurs between them that is related to the composition of the plates and the liquid. The best known example of the use of this phenomenon is the glass electrode used for measuring pH, but a range of ion specific electrodes are now available. A polarographic electrochemical method relies on oxygen diffusion through a Teflon membrane at ambient temperatures to produce a small current between two electrodes. Polarographic electrodes can be made as small as 2 mm in diameter, although chemical detectors are generally rather bulky and complicated.

9.2.1.4. Emerging transducer technology

Solid-state technology has led to the development over the last two decades of a number of miniature transducers suitable for the measurement of environmental conditions in foods. The output from these devices is not always as easily converted into a value on the appropriate scale as that from traditional transducers, but they can be combined with specialised ICs[2] to modify the signal to a more useful form, a process termed **signal conditioning**. For example, ICs can be designed to add to, multiply, linearise and even digitise transducer outputs. Such 'integrated transducers' remain extremely small, typically of the order of 10-100 mm^3. The development and potential of integrated transducers, also called smart sensors is reviewed by Brignell and Dorey (1983).

The temperature sensitivity of some semiconductor transistors, whilst undesirable in most applications, has been exploited to produce an ultralinear thermometer in the range -50 to 150°C. A microelectronic temperature sensor, the platinum resistance temperature detector (PRTD), is also available which yields very accurate measurements.

Several sensors for specific chemical species are now available based on solid-state detectors, namely specialised transistors called MOSFETs (Metal Oxide Silicon Field Effect Transistor). These can be made selective for different chemical species by the deposition of an appropriate polymer on part of the device. Sensors have been made that are able to measure CO, CO_2, CH_4, SO_2, and NH_3, as well as relative humidity. A pH meter based on this principle uses a polymer of Si_3N_4, Al_2O_3. A miniature dew point humidity sensor, based on capacitance changes due to condensation on an electrode and using a semiconductor temperature sensor and miniature Peltier cooling element (a thermocouple in reverse), has also been developed. A similar technique has been used in an oxygen sensor utilising the current generated in a mini electrochemical cell in response to the pressure of oxygen diffusing through a suitable membrane.

A new class of highly specific and sensitive sensor which may find applications in predictive microbiology is now also emerging. Biosensors utilise ion permeable membranes connected to microelectronic devices. While ions flow through channels in the membrane, an electrical current flows. Receptor molecules attached to the membrane can control the flow of ions through the channel; when the receptor binds to a target molecule the channel is 'blocked' and the current flow stops. The cessation of current flow may be detected by appropriate circuitry, and used as a signal to drive other devices and applications. The potential exists to make extremely sensitive specific receptors for molecules of interest.

A critical appraisal of these sensors is beyond the scope of this chapter; the above discussion is intended to demonstrate the scope of, and

[2] An IC (integrated circuit) is a very small solid state circuit which consists of interconnected semiconductor devices such as transistors, capacitors, and resistors printed into a single silicon chip.

potential for, the application of existing technology to predictive microbiology.

9.2.2. Timers

A variety of means for following the real and relative time course of processes is available. Monitoring of processes may be continuous, such as provided by a chart record, or at discrete points in time as occurs with loggers which record instantaneous measurements. In electronic devices, a clock circuit is employed based on an oscillating quartz crystal just as is used in most modern watches. The crystal oscillates extremely regularly when a voltage is applied to it, and conversely emits a voltage on the reverse cycle of its oscillation. This generates a regular voltage pulse which is used to synchronise the operations of the microprocessor, and can also be used to provide a real time record accurate to a few seconds per month.

9.2.3. Information Storage

The output of transducers can be recorded in a number of ways, but storage as digitised information on solid-state memory chips is most appropriate in all but a few specialised applications. Transducers produce an analogue signal, i.e. one which varies proportionally and continuously relative to the measured property. Analogue signals can be digitised, i.e. converted into a discrete number and encoded as a sequence of electric pulses which is proportional to the magnitude of the analogue value. Once digitised the output can be stored, transmitted or mathematically manipulated by computer. This conversion is undertaken by devices called, simply, analogue-to-digital (A/D) converters. There are a number of different types of A/D converters, but a discussion of them is beyond the scope of this text. It should be pointed out, however, that it is not always necessary to digitise transducer outputs in order to manipulate them. Analogue computers also exist which have the processing instructions built in to the electronic hardware, but such computers are usually dedicated to performing that one function. Predictive devices based only on analogue signals can be envisaged, as will be discussed later.

Digitised information is usually converted to binary form (expressed to base two), because binary encoded information can be transmitted and stored in the form of electrical pulses. Two types of memory circuit are available, one for permanent and one for temporary storage. Either of these types can store one piece of binary information called a 'bit'; the term 'byte' refers to a fixed number of bits generally corresponding to one character (i.e. a number or letter).

Random Access Memory (RAM), is used for the temporary storage of information such as collected data or data in the process of being manipulated. The fundamental storage unit in RAM is a 'bistable' or 'flip-flop' which is set to, and kept in, one of two voltage states as long as current flows in the circuit. Thus, a bistable codes for one bit. Bistables may be organised into 'registers' each of which contain one item of data, i.e. one byte. By using many bistables together it is possible to store large amounts of information. Thus electronic memory is somewhat like a simple black and white photograph. Each grain in the photographic emulsion can only take one of two states: black or white, but by using many of these points at once and in a specific order it is possible to reproduce a complex image. Similarly, by combining bistables into registers it is possible to record a byte of information, and by using even more to store a word or number. Large banks of bistables can be miniaturised on silicon chips; these form the basis of the memory of computers and other 'smart' electronic devices. Thus RAM requires the input of energy to maintain the data which is lost if there is a power failure.

Information in Read Only Memory (ROM) is entered once, in some devices indelibly, and then retained. ROM memory is based on diodes rather than bistables so that the information is an integral part of the hardware and is not lost when power is disconnected. ROM is used for the storage of permanent instructions which the device needs every time it is used,such as that for the management of the data, or its manipulation e.g. a mathematical model.

Different configurations of the bistable arrays determine the speed of processing, the size of the numbers that can be dealt with, and the memory capacity. It is the RAM capacity that limits the amount of information that the device can collect before transferring it to another storage device. RAM chips which can record up to 1 000 000 bytes are commercially available. The memory capacity of a device is often quoted as the numbers of bytes that it can store.

Implicit in the use of RAM is the need for a system to control and coordinate the orderly storage and retrieval of the information. This is one of the functions of the Central Processing Unit (CPU), the other of which is processing of the information according to its programmed instructions. These are the instructions found in ROM. In the late 1960's the technology became available to place as many as 40,000 circuit elements onto a single silicon chip, enough to construct a processor of reasonable flexibility, which could be programmed to perform an enormous range of tasks. It was reasoned that because this single electronic unit could be applied to so many tasks it could be produced in large quantities and, through economies of scale, inexpensively. The reasoning was sound; the device - a microprocessor - has become so cheap that it is often much cheaper to install a microprocessor, and use only a fraction of its capabilities, than to install a purpose built system. The Zilog Z80, a very widely used microprocessor, can be purchased for a few dollars. A microprocessor is the name given to a CPU on one or a few chips. It incorporates a fast acting arithmetic unit, a high speed memory and a control block that instructs the system what to do. Microprocessors can also receive new and supplementary instructions from an external source, accept and surrender data, and interact with other peripheral devices. Their generality of application makes them particularly useful in instrument systems because they can be programmed to condition transducer outputs. Output values can be corrected, as they are generated by a sensor, for defects such as non-linearity and drift from the true calibration. As well, they may be amplified and signals from different transducers equalised. These corrections may involve simple arithmetical application of a constant value, reference to tables of values in memory, or application of complex formulae. The term 'smart sensors' has been coined in reference to transducers in which intelligence is added by means of a digital processor, preferably on the same chip. A general circuit for smart sensors has been presented by Brignell and Dorey (1983) and is reproduced in Fig. 9.1, overleaf. The transducer is under total processor control, the processor containing full details of the transducer characteristics in ROM so that the final output is always an accurate gauge of the measured property. The actual type of transducer in Fig. 9.1 is also of little consequence.

The ability of microprocessors to perform complex manipulation of the data also makes them particularly useful for application as predictive food microbiology devices.

Although insignificant in terms of space, cost and power consumption, it is easy to see why microprocessors have become so important in the design of instruments. Coupled with compact memory chips and semiconductor transducers and smart sensors they provide the technology to construct compact environmental monitoring devices of relevance to the food industry.

9.2.4. Environmental Monitors/Loggers

A number of interests seem to be driving the development of better and less expensive sensors and monitors. For example, the automotive industry requires sensors for incorporation in engine management

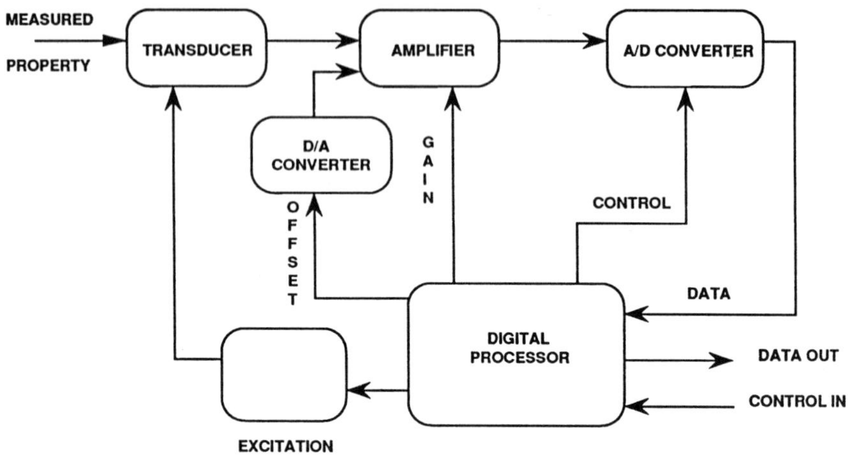

Fig. 9.1 Component parts of a typical 'smart sensor' of moderate complexity. The processor, in addition to transforming the data and managing its handling, may be used to control the offset and gain of the amplifier, the range of values accepted by the analogue to digital converter (for example, to exclude nonsensical values), or even the transducer itself. In this way, the signal from an imperfect transducer can be made to generate the desired response. Ideally, the entire circuit would be in a single housing and consequently invisible to the user. [*After* Brignell and Dorey, 1983].

systems, particularly temperature and (exhaust) gas detectors, whilst increased pressure for monitoring and control of industrial emissions has created another market. The food and pharmaceuticals industries are, of course, other potential users. Whatever the impetus, technological innovations have led to the commercial production of portable environmental loggers which can be applied to the needs of predictive microbiology, as well as devices dedicated for use within the food industry.

The term logger is used for a device which records some automatically measured parameter. Table 9.1, opposite, summarises the details of some

Table 9.1 **Summary of Some Temperature Loggers/Integrators Commercially Available**

Device	Temperature history interpretation as bacterial growth?	Sensor Temperature Range ψ (Min/Max °C)	Number of Channels	Total Record Capacity	Direct Readouts on Device	Operational Life Without Service or Battery Replacement	Approx. Cost ($US)	Manufacturer
Don Whitley TTFI	yes	as needed	1	- - -	Shelf Life Used	years	700	DON WHITLEY SCIENTIFIC LTD., West Yorks., UK.
Delphi	yes	-20/+40	1	8,000	none	5 years	1000	MANAGEMENT INFORMATION RESOURCES LTD., Wellington NZ.
Smartlog	yes	-10/+39 or as needed	1	32,000	THI*	100 days	900	REMONSYS, Keynsham, Bristol, UK.
Autolog	no	various (-29/+150)	4	5,124	device status	1 years	1000	" " " "
Control One	no	various (-40/+105)	2	250,000	none	10 years	1000	CONTROL ONE, Old Greenwich, Connecticut, USA.
Thinline Datacorder	no	-30/+30	>10?	- - -	none	60 days	- - -	CARRIER TRANSICOLD, Syracuse, NY, USA.
Squirrel	no	various (-50/+150)	16	32,000	temp.	6 months	- - -	GRANT INSTRUMENTS, Barrington, Cambridge, UK.
Thermoking	no	- - -		- - -	device status	80 days	- - -	THERMOKING CORP., Minneapolis, USA.
Anritsu	no	various (-100/+1200)	1-6	3,000	temp.	days	2000	ANRITSU Meter Co., Ltd., Meguro-ku, Tokyo, JAPAN.
DTR(disposable)	no	-25/+35	1	†	temp history	1-8 weeks	25	DTR Co., Ltd., Modesto, California, USA.
Datataker	no	various (-250/+1800)	10-50	300,000	none	18 months	- - -	DATA ELECTRONICS, Rowville, Vic., AUS.
Auscord	no	various (-50/+1000)	1	†	temp history	20 hours	1000	FOSS ELECTRIC, Terry Hills, NSW, AUS.
ACR	no	various (-40/+125)	1-8	32 000	none	10 years	7-1200	ACR Systems, Inc., British Columbia, CANADA.

ψ The operational range of most devices' recording unit is ≈ -40 to +70°C; external sensors allow wider ranges to be monitored.
† These devices store the information as a continuous paper chart record.
* THI = 'temperature history index' See text for details.

of the loggers that are available at the time of writing. The list is not exhaustive, but limited to those devices for which technical details were available to us, and which appear to have some relevance to prediction of microbial growth in food. Not all of the devices were originally intended for use in the food industry, but are included to demonstrate the potential of existing commercial logging devices.

The devices each have their own intended application which partly determines the design of the device and its characteristics. The differences include :

i) the types of transducer available and their ranges, sensitivities, and accuracy;
ii) memory capacity;
iii) environmental limits for reliable operation, and general robustness;
iv) battery life;
v) number of channels which can be monitored and the number of different *types* of transducer output that can be logged;
vi) size and weight;
vii) security, i.e. whether the device can be protected against unauthorised operators;
viii) whether the device is self-contained in relation to setting up and data retrieval and analysis, or must be connected to a computer or other device for these operations;
ix) complexity of data manipulation and analysis available; operation in real time;
x) frequency of measurement and whether it is user-selected.

As might be expected, all these 'features' come at a cost and so a decision must be made regarding the real requirements of a predictive device so that it is a cost-effective quality assurance tool. Furthermore, the same capabilities may not be required for all types of food and all types of storage and it may be more appropriate to think in terms of a range of devices each designed to perform a particular task.

For example, milk is often stored and transported for up to one or two days prior to its arrival at the milk processing plant. To assure its storage life milk should be maintained in the chill range during that time and it is in the interests of both the farmer and the processor to know the temperature history of the milk during that time also. It would be a simple matter to require farmers and transporters to include a temperature logger with all shipments of milk, but it would require considerable capital outlay, first of all to purchase all the loggers needed, and then to supervise and coordinate their use. The DTR is a disposable device which consists of a bimetallic thermostatic coil, accurate to $\pm$ 1°C over the range -25 to +35°C, connected to a pen which moves across a spring-driven paper chart recorder. It is intended to travel with the milk, and can be examined upon receipt at the factory for excursions beyond acceptable temperature limits and then discarded. The units, if purchased in bulk, cost ≈ US $20 each

The devices listed in Table 9.1 all have the capacity to monitor and record the environmental information necessary to enable a prediction to be made regarding the microbiological quality of a food. However, only a handful include the progamming necessary to make such predictions. To do this they must be able to add up the effect of the conditions experienced at each stage of the product's life on the growth of microorganisms, a process known in mathematics as **integration.**

9.3. INTEGRATION

9.3.1. Background

Predictive microbiology relies on the prediction of rates of growth of microorganisms. The rate is expressed in units of generation per unit time (e.g. gen h^{-1}). Ultimately our interest is in how this relates to actual numbers of microoorganisms on the food and its consequences for safety and remaining shelf life, i.e. we need to know how many generations of growth have been possible in the period monitored. If we assume that, under the conditions of interest, the time taken for growth rate to 'equilibrate' to new conditions is negligible, the answer is given by the product of rate and time:

$$\text{Growth rate (gen h}^{-1}\text{) x Time (h)} \\ = \text{Number of generations}$$

Note that the units of measurement are physically consistent. The number of generations corresponds to the number of population doublings, thus three and a half generations represents a greater than tenfold increase in microbial numbers, and six and a half generations correspond to nearly a one hundred-fold increase.

Geometrically, integration of rate can be visualised as the area under a graph of rate *versus* time. Provided the growth rate is constant the above calculation is a simple one, as is its geometrical interpretation (Fig. 9.2a, overleaf).

Were constant conditions possible there would be no need for loggers and predictive microbiology. Nonetheless, where rate fluctuates the same

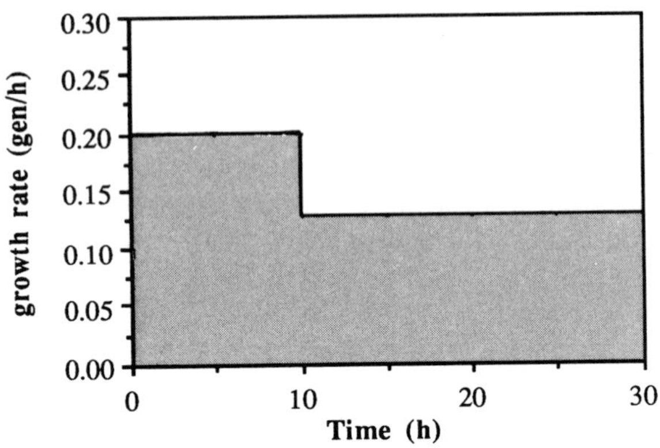

Fig. 9.2a The figure shows the growth rate and its dependence on temperature. The number of generations predicted over the temperature history is given by the product of rate and time. In this example, the temperature fluctuates only once and the calculation is simple, i.e.

Number of generations $= 10\text{ h} \times 0.20\text{ gen h}^{-1} + 20\text{ h} \times 0.13\text{ gen h}^{-1}$
$= 2 + 2.6$ generations
$= 4.6$ generations

which is equivalent to an ~25-fold increase in the population density.

mathematics hold; they are simply more difficult to apply. The solution is still to determine the area under the rate-time graph and there are a number of ways of doing this. A primitive method would be to plot the data on graph paper and count the squares under the curve, or even to cut out the area under the curve and weigh it or otherwise measure its area. A more accurate method is to divide the graph into many small time intervals and determine the number of generations in each using as an approximation the average rate value in that interval. The growth calculated for each of these intervals is then added to give the total number of generations over the entire time (Fig. 9.2b). The smaller the time interval chosen the more accurate will be the estimate. If the relationship between time and rate can be described by a mathematical equation, the time interval can be chosen to be vanishingly small and an

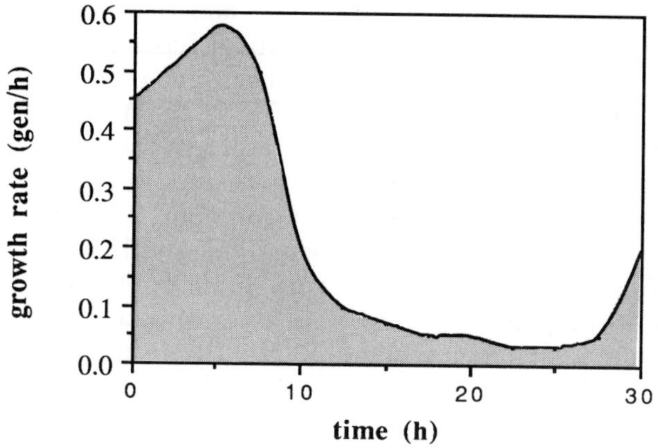

Fig. 9.2b In this example the temperature fluctuates rapidly and continuously and the calculation is made more difficult. However, the same approach may still be used to estimate the number of generations by dividing the temperature/rate history into many small time intervals as depicted on the figure and summing the growth predicted in each. As the temperature/growth-rate is continually fluctuating we can approximate the amount of growth in each interval by multiplying the time period by the average of the fastest and slowest rate in that period. The shorter the time interval chosen the more closely will the estimate approximate the true value.

exact value for the area under the graph calculated. This is the mathematical process of integration.

9.3.2. Temperature Function Integration and Food

Temperature integration as a means of assessing storage conditions is not a new concept. It was applied to fisheries products by Nixon (1971), who based his work on results showing that the relative spoilage rate of fish over the temperature range 0-10°C and above a certain limiting value (analogous to T_{min}), was directly proportional to temperature. This enabled him to integrate temperature values directly and then multiply the result by a constant value to obtain the extent of spoilage relative to that known at some reference temperature.

The simple relationship between temperature and relative rate of spoilage used by Nixon (1971) is inadequate to describe rate of growth or spoilage over a wider temperature range and for other systems, and we now know that more complex mathematical functions are required to model the relationship between temperature, and other environmental factors, and microbial growth rate. The first step in an environmental history integration would be to use the predictive model to convert the values of the measured environmental parameters at each time interval into a predicted growth rate. The environmental factors governing microbial growth rate in food are not usually a simple function of time, unless the product is stored under very well controlled conditions. Accordingly, we cannot express the environmental history as a mathematical function to be integrated implicitly. Instead we must revert to first principles, a tedious, repetitive process requiring the determination of a mean value of rate over the time interval, its multiplication by the time interval, and the summation of all the separate intervals. Alternatively, one could take a mean of all the rate values calculated and multiply this by the total time.

An ingenious device to simplify the process was devised by Ronsivalli and Charm (1975) and is shown diagrammatically in Fig. 9.3a, opposite. This mechanical integrator called the ‘shelf life prediction slide rule’, or ‘predictor’, consists of three parts:

i) an index (sliding cursor),
ii) a shelf life scale, and
iii) a temperature scale.

The user selects the observed temperature on the temperature scale, and moves the cursor along that line for the appropriate amount of time that the product has been at that temperature. Starting with the cursor at this new point, the next time-temperature combination is added. This process is repeated for each time-temperature combination the product is known to have experienced (for example, by reference to a chart record), so that the cursor always moves further to the right. Once all parts of the temperature history have been integrated, the overall effect on storage life is determined by reading the final position of the cursor against the shelf life index, which represents the remaining shelf life of the product at some ideal reference temperature. One of the claimed benefits of the device is that users can see the economic benefits which can be realised by keeping the product at as low a temperature as possible. At the time of invention the relationship between temperature

a.

b.

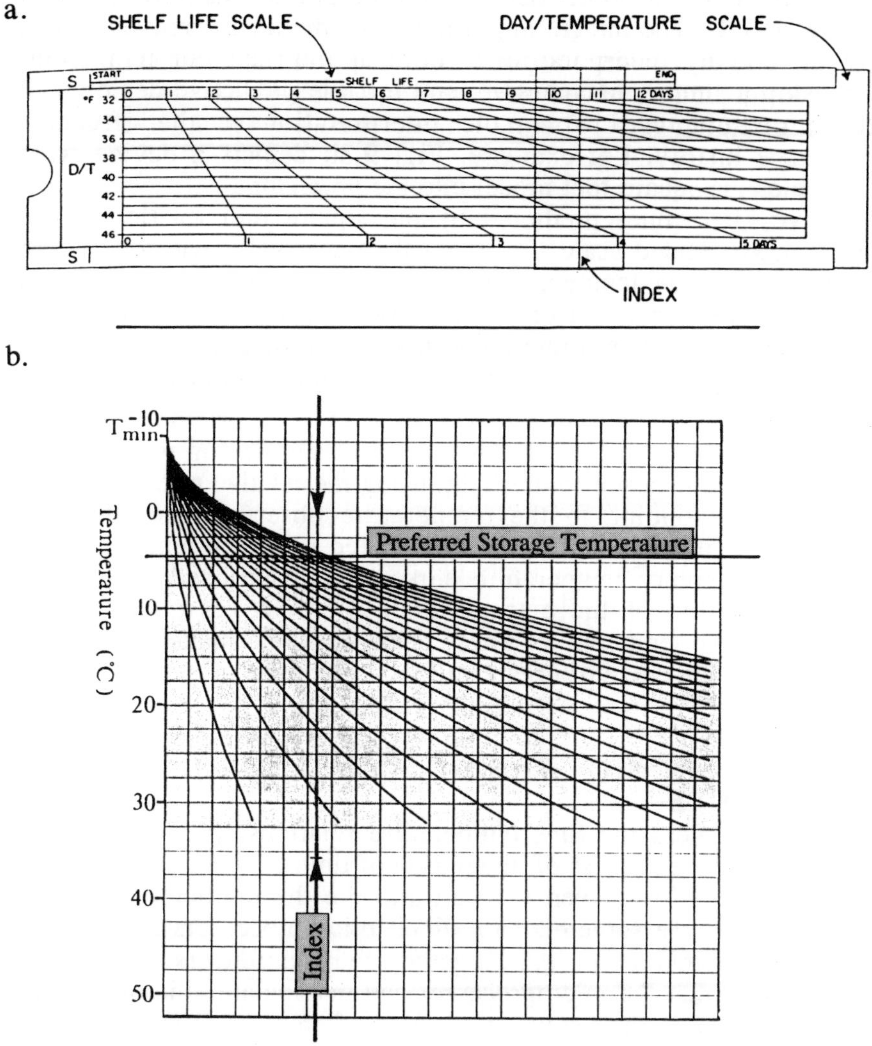

Fig. 9.3 'Low-tech' time temperature integrators. a) The shelf life prediction slide rule, or 'predictor'. [Reproduced from Ronsivalli and Charm (1975) by permission of Natl. Oceanic and Atmospheric Admin., U.S. Dept. Commerce, Washington]. b) A modification of the 'predictor' with more general application, and based on the square root model (Eqn. 3.10). The operation of these devices is explained in the text opposite.

and spoilage rate was poorly defined and the device was only reliable in the range 0-7°C. The integrating principle of the device remains sound, however, and by incorporating a more appropriate rate-temperature relationship a simple, inexpensive, and reliable predictive device could still be manufactured to enable interpretation of the effects of temperature history. For example, Fig. 9.3b depicts a device which incorporates the square root model into the predictor.

Relative rate values predicted by the square root relationship are shown on the shaded scale, for equal time intervals. The relative rate scale slides along the temperature axis, however, so that the predictor can be used for any organism or spoilage process which the square root model describes. The zero rate point, marked 'T_{min}', is aligned with the temperature corresponding to the T_{min} of the organism or spoilage process being integrated. The cursor is used as in the original version; i.e the growth at each time interval is integrated by moving the cursor the appropriate time measure to the right, using the relative rate scale, along the line corresponding to the temperature observed over that time interval. In addition, rather than using a fixed reference temperature, an adjustable 'reference temperature cursor' could be used. This would enable calculation of the elapsed shelf life relative to that at *any* preferred storage temperature. These changes would make its operation general for a wide range of organism/food systems.

9.4. ELECTRONIC PREDICTIVE MICROBIOLOGY DEVICES

To reiterate, unless the time-temperature relationship can be defined mathematically the process of integration of time-temperature histories is tedious and repetitive, even with an aid like the predictor. Repetitive tasks, however, are something to which microprocessors are perfectly suited. It is for this reason that, together with transducers to supply the data in a form that the microprocessor can manipulate, microprocessors hold one solution to the routine application of predictive microbiology.

We have demonstrated that the technology exists to monitor and record the magnitude of many of the environmental factors governing the rate of microbial growth on foods. We have also shown that mathematical models can be used to interpret these parameters in terms of microbial growth rate. Square root models exist for temperature and water activity (Chandler and McMeekin, 1989a; McMeekin *et al.*, 1987; Ross and McMeekin, 1991), as do Arrhenius-based models (Broughall *et al.*, 1983, Davey, 1989b). Polynomial models incorporating temperature, water activity, NaCl concentration, nitrate concentration, and aerobic/anaerobic conditions exist also for a number of recognised food-borne pathogens (Buchanan and Phillips, 1990; Gibson *et al.*, 1988). Microprocessors can be programmed to integrate the predicted rate over time, and this information used to predict the remaining shelf life of the product or its microbiological safety.

9.4.1. Early Devices

Circuitry for one type of electronic predictive microbiology device, namely a temperature function integrator, was presented by Nixon (1977). In his patent application two devices were described which monitor temperature with either an NTC thermistor or silicon-based semiconductor diode, condition the signal from these transducers in accordance with a temperature function appropriate to the spoilage of fish, and integrate that temperature history over time. In addition, Nixon's designs incorporate transmitters so that the temperature can be remotely monitored, and assessment made at any time without having to remove the device from its location. The first device uses analogue processing of a temperature signal (electrical current) which has an instantaneous value directly proportional to the rate of spoilage. Integration is achieved by using the current to control the rate of a pulse generator which drives a digital counter. This design has been incorporated into a device called the Tefimupot (temperature function integrator multi-point telemeter) intended to enable use of multiple sensors in fish holds with telemetry delivering the information to a central control station. The versatility of analogue circuits to condition signals is evident from data of Nixon, presented by Olley and Ratkowsky (1973b), showing the ability of various circuit designs to simulate the temperature-fish spoilage response.

The second device uses digitised information which "enables the flexibility and accuracy of the temperature function integrator to be considerably increased, since the signal modifier can take the form of a PROM[3] which allows the temperature-rate of deterioration relationship to be as accurate as food technology data permits" (Nixon, 1977). The basic principle of this device is that for each temperature able to be discriminated[4], there is a unique rate-of-spoilage value held in the PROM. Temperature signals are thus converted to corresponding rates

[3] Programmable Read Only Memory

[4] The temperature signal has been digitised; i.e. there is a limited number of discrete temperature values which can be recorded.

which are integrated over time. Nixon further points out that, if necessary, a separate PROM chip could be employed for different food/microbe combinations.

At about the same time (late 1970's) a number of simpler physical temperature function integrators were also being developed as cost-competitive alternatives to chemical or biochemical colour change indicators which generally do not accurately mimic the temperature-rate of spoilage relationship in most foods (see Chapter 8). Physical devices were thought to have greater potential because the temperature-rate function could be dictated by circuit design. These devices were based on microcoulometers (Storey and Owen, unpublished). The principle of operation is that temperature is converted into an electrical signal proportional to the rate of deterioration. The microcoulometer measures the total current which has flowed and in this way an integral of the temperature (rate) over time is obtained.

The microcoulometer output can be visualised in a number of ways. One microcoulometer device, known as an 'E-cell', comprises a silver case with a centre electrode of gold, the intervening space being filled with an electrolyte. As current flows through the cell, silver is deposited on the centre electrode in direct proportion to the current flow. i.e. the amount of silver deposited increases with the magnitude of the current and time. Including a thermistor in the circuit regulates the current flow in proportion to temperature so that the amount of silver deposited becomes a measure of the product of temperature and time, i.e. a time-temperature integrator. The 'E-cell' is read back by reversing the direction of current flow so that the silver is removed. The current is supplied at a constant rate and the time taken to remove the silver recorded, which yields a value for the time-temperature history (see Carseldine and Weste, 1976).

There are a number of technical limitations of the microcoulometer devices but the main objection, in hindsight, is that an inappropriate temperature function was used to model the rate of product deterioration.

Storey and Owen (unpublished), however, described a device based on the Ratkowsky *et al.* (1982) [i.e. square root] temperature function. The device relied on the non-linear change in resistance of a thermistor to drive a voltage-controlled oscillator, the output of which is integrated and displayed using an IC clock and liquid crystal display. This device was subsequently refined (Owen and Nesbitt, 1984) by employing erasable programmable ROM (EPROM) chips in a similar manner to Nixon's second device, to give greater accuracy than the earlier analogue device.

9.4.2. Contemporary Devices

At the time of writing, three commercially available electronic devices can be regarded as true predictive systems. These are:

i) the 'Don Whitley Time Temperature Function Integrator'
ii) the Remonsys 'Smartlog'
iii) the 'Delphi' temperature logger

9.4.2.1. The 'Don Whitley Time Temperature Function Integrator'

The Don Whitley Time Temperature Function Integrator was developed by Owen from the Owen and Nesbitt (1984) integrator. It is in principle an electronic clock which runs at real time at the optimum storage temperature for the product (e.g. 0°C for fresh fish products). At temperatures above this reference temperature the clock runs more quickly, while at those below it runs more slowly. The relationship between the temperature and the clock's rate is the same as that of bacterial growth rate response to temperature. The output of the device is a continuous display of 'elapsed storage life': because the expected shelf life of the product at the reference temperature is known, the display shows how much of the shelf life is 'used up' and how much longer it can be kept at the reference temperature before spoiling. The rate at which the clock is running is also a good visual index of the effect on product quality of the temperature of different processing steps. A schematic diagram of the device is given in Fig. 9.4, overleaf. The device relies on the square root relationship in its programming to generate a rate value at each temperature relative to the spoilage rate at the optimal storage temperature at which the normal shelf life is known. For example, assuming that spoilage is due to psychrotrophic organisms, which typically have a T_{min} of ≈ -10°C (263 K), the relative rate at any temperature T (°C) within the suboptimal temperature range to that at 0°C is given by:

$$\text{Relative rate} (T) = \frac{(T - -10)}{(0 - -10)}$$

$$= \frac{T}{10} + \frac{10}{10}$$

$$= 0.1\, T + 1$$

Note that this equation would change for products where the reference temperature is other than 0°C, or where the T_{min} of the organisms of concern is other than -10°C, as shown in Table 6.2.

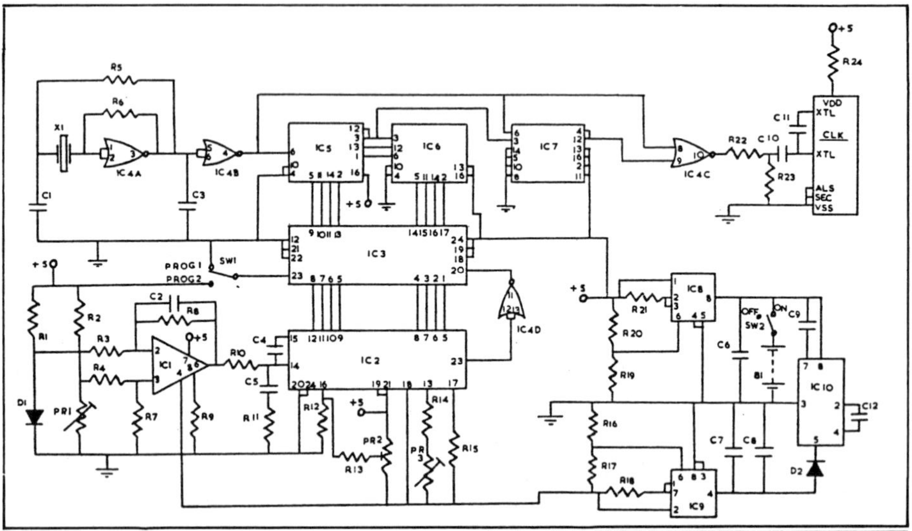

Component values for integrator (Figure 2)			
C 1, 11, 12	100p	C5	300p
C 2, 8, 9	47n	C6, 7	100µ 25V
C 3	500p	C10	47p
C 4	75p		
D 1, 2	IN 4148		
IC1	LM4250	IC5, 6, 7	HEF 4526B
IC2	8703 CJ	IC 8	ICL 7663 CPA
IC 3	IM 6654 MJG	IC9	ICL 7664 CPA
IC 4	MC14001B	IC10	ICL 7660 CPA
R 1, 22	100K	R12	10K
R 2, 5	47K	R 13, 24	270K
R 3, 4, 7, 15	1M	R 14, 16, 19	200K
R 6	10 M	R 17, 20	560 K
R8	4 M 7	R 18	39 R
R 9	8 M 2	R 21	68 R
R 10	27 K	R 23	22 K
R 11	100 R		
PR 1	20 K preset	PR 3	100K preset
PR 2	200 K preset		
SW1	SPDT		
SW2	SPST		
CLK	Clock module 304-841		
X1	32 KHz crystal from clock		
B1	5 × MN 1500		

Fig. 9.4 Circuit diagram, and legend for component values, of the Don Whitley Time Temperature Function Integrator. [Reproduced from Owen and Nesbitt (1984) with permission of United Trade Press, London].

The corresponding relative rate values derived from the above equation for each of the discrete temperatures in the selected range are converted to a 'division ratio', which is then stored at the appropriate memory locations in the EPROM.

The signal from the temperature transducer is converted to an eight bit binary number. This binary number is fed to the EPROM, and releases from the appropriate memory location for that temperature the corresponding divisor ratio value. The divisor ratio (analogous to relative rate) values from the EPROM are used to govern the number of pulses sent to the clock chip and display. The use of a set of values of rate in an EPROM to condition the temperature signal is known as a 'look-up table'. Relying on a look-up table imposes a number of restrictions on the device, for example, because the device is eight bit, there are only 256 possible memory locations. This limits the temperature range to 256 discrete values, which at 0.1°C accuracy limits the predictive ability of the device to a range of 25.5°C.

The TTFI uses a standard IC clock chip and liquid crystal display which shows hours and minutes. The storage life at the reference temperature however will usually be many days. Owen and Nesbitt (1984) chose to retain this display but to use the 'hours' display for days, and the 'minutes' display for fractions of days. As a result the output, which is the equivalent time elapsed at the reference temperature, is expressed in days and 1/60th of days. They point out that this need not cause confusion because, at a first approximation, 1/60th of a day corresponds to about 30 minutes. The same display is retained in the TTFI.

9.4.2.2. The Smartlog

The Smartlog is a temperature logger which also displays an indication of the product's storage conditions, the "Temperature History Index" (THI) defined as:

$$\text{THI} = \frac{\text{Elapsed Shelf Life}}{\text{Elapsed Time}} \times 100$$

A THI value greater than 100 indicates that the product has been subjected to temperatures which overall have shortened its expected shelf life relative to that at the reference storage temperature, i.e. it has been subjected to less than favourable temperature conditions. Conversely, a THI less than 100 indicates that the product has been handled under better than expected conditions and will achieve a longer than expected shelf life if subsequently kept at the reference temperature. The elapsed shelf life is calculated by the Smartlog using a predictive microbiology model, and also allows the user to set and reset the reference storage temperature to suit different products. The calculation is based on a square root model and is similar to that employed by the Don Whitley TTFI. The information processing is different, however, in that floating point arithmetic is used rather than a look-up table to calculate the result. The actual calculation of the THI factor is:

$$\text{THI factor} = \frac{(0.1 \times T \text{ actual} + 1)}{(0.1 \times T \text{ reference} + 1)}$$

(T. Lewis, *pers. comm.*)

It can be seen that this relationship is simply the quotient of the relative rate relationships for psychrotrophic spoilage organisms used in the TTFI. The THI factor assumes that T_{min} is constant and it cannot be altered by the user, though if necessary other T_{min} values could be employed in the Smartlog. The THI factor is multiplied by 100 to yield the THI.

The real elapsed time is also displayed which allows estimation of the elapsed shelf life at the ideal temperature. The shelf life index calculation is claimed only to be valid over the range -3 to 15°C.

The Smartlog can also act as a logger and is able to record up to 32,000 temperature values using an onboard thermistor accurate to ±0.2°C, and at user selected intervals ranging from 1 to 60 minutes. Battery life is greater than 100 days.

9.4.2.3. The Delphi temperature logger

The Delphi Temperature Logger and accompanying software provide an alternative, more detailed, assessment of the effects of temperature history on product quality. The logger is a compact, sealed unit powered by a lithium cell with expected battery life of five years. It can record up to 8000, eight bit logs. It has ± 0.25°C accuracy over its recording range (-20 to +40°C), though other ranges and resolutions are available on request. Logging intervals can be set from ~2 min to ~8 hours and a start delay of up to 85 days. A four character password provides data security. The temperature sensor is mounted within a Teflon or stainless steel probe, depending upon the type of sample it is required to monitor.

Computer software written for some of the loggers mentioned in Table 9.1 is available, but in most cases is limited to analyses of the temperature values. For predictive microbiology purposes it is necessary first to relate the measured parameters to microbial growth parameters before undertaking data analyses.

Software has been developed for use with the Delphi which enables both shelf life prediction and assessment of the hygienic efficiency of processing operations of red meat products. The basic element of the software is a routine which integrates the temperature history by essentially the same 'longhand technique' described earlier, i.e. by considering the temperature over discrete time intervals, applying a temperature-rate function to the temperature data, multiplying that rate by the time interval, and summing the growth predicted over each interval to yield an estimate of the growth over the entire temperature history. This type of program is not unique to the Delphi logger. A program available from Grant Instruments Ltd., (Barrington, Cambridge, UK) was used earlier by the MIRINZ group (Gill and Harrison, 1985) to integrate temperature histories obtained with a Squirrel digital meter/logger. In this program, growth was calculated within sequential time intervals from the growth rate for the average temperature within each interval, and the total growth obtained by summation.

9.4.3. Applications Software and Expert Systems

9.4.3.1.Shelf life prediction

The Delphi software for shelf life prediction (Gill *et al.*, 1988a) attempts to provide the specialised knowledge necessary for buyers to make an assessment of the remaining storage life of chilled red meats packaged under vacuum or CO_2 enriched atmospheres after shipment to distant markets. The program relies on models based on lag resolution times and growth rates for psychrotrophic enterobacteria in response to temperature, and anaerobic and high CO_2 conditions. It also requires an 'assured storage life' value for the product if the product is other than one of those for which assured storage life values are held in the logger memory. In addition to the assured storage life value, the program requires the temperature history from the logger before the analysis can begin. The program operator must input consignment identification number details and set start and stop times for the temperature history as the log will often contain records prior, and subsequent, to the time actually spent with the product.

The temperature history is first scanned for excursions above 7°C, and the potential for *E. coli* growth during any such periods is assessed by reference to a model specific for its growth in red meat products which is discussed below. If *E. coli* proliferation is calculated to exceed three generations, a possible health hazard is indicated. The entire temperature history is analysed to determine the growth of enterobacteria taking both lag and growth phases and packaging into account. The initial numbers on the product are estimated from the assured storage life value using the lag period and growth rate values appropriate to the logged temperatures. This initial microbial load estimate is then added to the calculated growth over the temperature history to yield an estimate of the absolute numbers on the product at the end of the log. The difference between this value and the assumed spoilage level of 10^7 organisms cm^{-2} defines the remaining storage life at any given temperature. It is apparent from the earlier discussion on relative rate calculations that this program contains some redundant steps; given that an assured storage life value is available, and an appropriate temperature rate model, there is no need to calculate absolute numbers to estimate the remaining shelf life at any chosen temperature.

9.4.3.2. Hygienic efficiency of meat processing

The program for evaluating the hygienic efficiency of meat processing procedures (Gill *et al.*, 1988b) is based on calculation of the greatest possible amount of growth of *E. coli*, as a model of pathogen growth responses, on meat subject to that process. The process assessment software is interactive and initially requests from the user details of the type of process. It then prints a form to accompany the logger through the processing operations. The form provides instructions for placement of the probe and logger within different product types and space for recording times of insertion and removal of temperature probes at stages of processing in which it is not possible for the probe to remain within the product. This information is subsequently used in the program to edit out those periods in the temperature record not representative of product temperature. Upon completion of the process the logger is again linked to the computer, a time-temperature graph displayed, and the probe insertion/removal times entered by the operator. The software checks that the temperature history data is consistent with the nominated process. If it is, analysis proceeds; if not, an error message is printed requesting the operator to resolve the discrepancy. From the process description the program selects the appropriate data sets from the model to apply to the particular stages of processing. Data sets are included for the resolution of lag phase at temperatures permitting growth after chill and frozen storage, under aerobic and anaerobic conditions, and for variation of growth rate with temperature. At stages where logging was not possible within the product, the program assumes that the higher of either the starting or final temperature of those unlogged periods applied throughout. Finally, the process description, temperature history and analysis of potential *E. coli* growth are printed.

9.4.3.3. The 'Pathogen Modeling Program'

The 'Pathogen Modeling Program' (Buchanan, 1991) embodies a different concept to the above devices. The program is in fact a set of databases accessed via the spreadsheet component of [5]Lotus 1-2-3™ software and is proposed for use in developing food products of enhanced microbiological safety and for illustrating the multivariate nature of microbial growth on food. The commercial spreadsheet software was chosen because of its highly visual nature, array of useful mathematical functions, ready accessibility of good graphics and simplicity of developing menu-driven option selection protocols all of which combine to make the final program 'user-friendly'. The particular software chosen is also widely used. There is no reason, however, why the same style of program could not be set up within another commercial spreadsheet package which has similar capabilities.

The 'Pathogen Modeling Program' software incorporates models drawn both from the literature, and work undertaken at the USDA Eastern Regional Laboratories, for the growth of at least seven pathogenic bacteria in response to some or all of the following parameters: temperature, pH, NaCl concentration, sodium nitrite and aerobic/ anaerobic atmosphere. The software uses various macro commands available within the software to speed up and 'hide' from the user the calculations being undertaken by the spreadsheet software. As described in Chapter 3 the USDA group model the natural logarithm of 'modified Gompertz' growth curve parameters with polynomial expressions. From these parameters times to reach specified microbial loads, lag times and generation times can be calculated, and it is these values that are generated by the software. No attempt is made to integrate the effects of fluctuating storage conditions over time, and consequently no use is made of environmental monitoring devices and loggers. Furthermore, because the data are derived from growth of pure cultures the predicted values may differ from those actually observed in real food

[5] Lotus 1-2-3™: Lotus Development Corp. 55 Cambridge Parkway, Cambridge, Massachusetts, USA.

products, which is a particular advantage of the use of relative rates or assured storage life values.

9.4.3.4. Other databases and 'expert systems'

A number of other groups are working toward the development of software and systems incorporating predictive kinetic models. In the UK, the United Kingdom Predictive Food Microbiology Database (UKPFMD) is being established with MAFF funds. The system, depicted diagrammatically in Fig. 9.5 opposite, currently involves a database of kinetic results, and a library of predictive models which have been approved for use.

Users access the data and models bases via 'expert centres', where trained staff, familiar with modelling, use the system and provide interpretation of the results generated. In addition, work has recently started on an 'expert system', which is intended to embody the expertise of experienced microbiologists and modellers, so that non-expert users can access the databases directly. The expert system will request from the user the information necessary to determine the appropriate model and produce the required prediction (Walker and Jones, 1992; Jones, *in press*).

The Unilever Company has already established a food microbiology expert system. The system incorporates a set of rules based on the knowledge and 'decision-making' processes of an expert food microbiologist, as well as a database of knowledge of chilled food manufacturing and food-borne microorganisms, and is linked to a spreadsheet providing predictions based on microbiological growth models (Adair *et al.*, *in press*).

In Canada, software is being developed to assess product safety and to define production system critical control points. MKES (an acronym of Microbial Kinetics Expert System) Tools simulates the growth and survival of pathogenic organisms when subjected to many different environmental conditions. The responses predicted are used to estimate the significance of the individual factors in that situation (Voyer, *in press*).

Similarly, Zwietering *et al.* (*in press*) have reported on the development of software which models bacterial growth in food production and distribution chains.

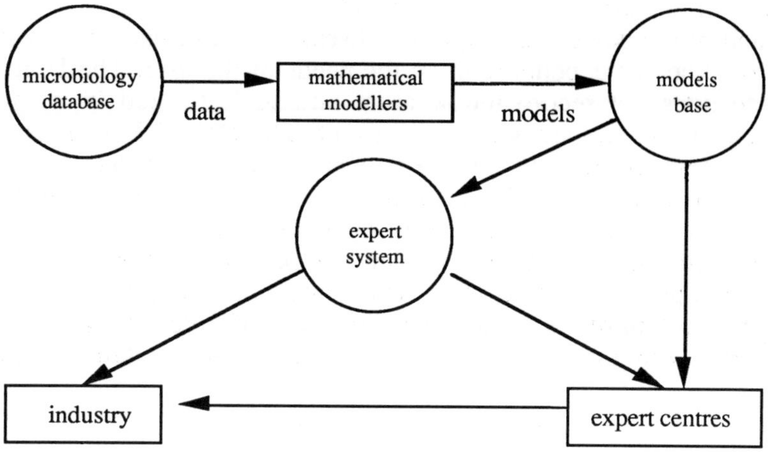

Fig. 9.5 Diagram showing the interrelations of the United Kingdom Predictive Food Microbiology Database. [*After* Walker and Jones, (1992)]

9.4.4. Conclusions

Theoretically any of the electronic temperature loggers could be combined with predictive microbiology computer programs to yield estimates of the extent of growth of any microorganism of interest. Software can be written to analyse an environmental history, or any part of it, and using a predictive microbiology model relate this to the specific increase in the number of bacteria of interest, whether spoilage organisms or pathogens. A relationship of any complexity could be incorporated into a predictive device, in principle, by recourse to a look-up table of experimentally determined rate values included in the device's ROM. Consequently, it is not strictly necessary to model mathematically the environment-growth rate relationship. A single software package could include models for the growth of different bacteria on different foods. Using such information it is theoretically possible to assess objectively the effect of any given period of the environmental history on the overall shelf life and safety of a product. This information would be invaluable for Hazard Analysis, appraisal of problem areas in processing and distribution chains, for allocation of responsibility for loss of product quality through mishandling and for evaluating cost *versus* benefit of changed processing protocols.

Each of the devices and systems discussed above has its own strengths and weaknesses depending upon the needs of the user. The Don Whitley device does not record temperature and so is limited in its ability to pinpoint problems areas after the event. The Delphi Logger is intended to be tamper-proof and consequently has no display; it must be connected to a computer before the information can be retrieved and analysed. Whilst the Smartlog immediately displays an index of the temperature history and also collects a complete temperature history it does not yet have computer software to translate the temperature history into the increase in numbers of specific bacteria of interest. The 'Pathogen Modeling Program' is limited to prediction for constant conditions of storage.

9.5. SUMMARY

The compactness, lower power consumption, robustness, comparatively low cost and 'programmability' of semiconductor technology are such a powerful combination that it is unlikely that any new predictive device would use older technology. The technology already exists to develop devices able to monitor many of the environmental parameters in foods which affect the rate of microbial growth and thus, using microprocessor technology, to yield estimates of food quality and safety.

Realisation of the full potential of this technology will rely upon the development of cost-effective transducers and their combination with information processing devices, and on derivation of reliable models for bacterial growth kinetics which can be integrated into such devices. The commercial availability of a number of predictive devices indicates that this is a technology that can be practically employed in the food industry and other industries concerned with perishable commodities.

10.

A THERMODYNAMIC APPROACH TO BACTERIAL GROWTH

The square root model has been shown to model bacterial growth well but has been criticised for lacking a mechanistic basis. Using as a starting point the assumption that bacterial growth is determined by a single rate limiting enzyme, the concentration of which is determined by inactivation due to temperature, a new mechanistic model is derived. This model, consistent with contemporary paradigms of heat and cold denaturation of proteins, is based upon the synthesis of previously published thermodynamic models. Predictions from the new model were generated using realistic values for the thermodynamic parameters. When presented as a square root plot, the predicted rate-temperature responses suggest a possible explanation for the form of the square root model.

10.1. INTRODUCTION

"If one surveys what is known about microbial physiology, one almost immediately notices a strong dichotomy. On the one hand, much is known (empirically) about the growth rates and rates of substrate consumption for different microorganisms growing on various substrates. On the other hand, the biochemistry and molecular biology of the organisms are known in considerable detail. However, between these two well understood aspects of microbial growth there is a domain where our understanding is far less complete: this concerns the question of how the growth kinetics of microorganisms are related to their biochemistry" (van Dam *et al.*, 1988).

10.1.1. The Trouble with Empirical Models....

The square root model must be considered to be empirical, and has drawn some implicit criticism in consequence of this (Heitzer *et al.*, 1991). It is interesting to speculate nonetheless on *why* the relationship between the rate of biological processes and temperature is well-described by this model. In this chapter we engage in a 'flight of fancy'; we attempt to develop a mechanistic temperature-growth rate model encompassing contemporary knowledge of the effects of temperature and protein solvation properties on enzyme activity, as envisaged in the pioneering work of Brandts (1964-1969).

Thus we have moved from a largely descriptive perspective, in which observations were made and explanations sought, to a prospective approach in which we develop hypotheses which could drive the design and direction of future experiments. We did not know where this exercise would ultimately lead, but remained alert for a possible convergence between the empirical square-root model and the mechanistic thermodynamic approach.

10.1.2. Model Basis

Despite the complexity of the interrelationships of physiological factors affecting growth rate, a common starting point in the development of mechanistic models relating temperature to rates of biological processes has been the assumption that growth rate is determined by a single, key enzyme which is rate-limiting at all temperatures, and the effective concentration of which is affected by inactivation at both high and low temperatures. Despite criticism of this approach, both overt (see Dawes, 1964) and implicit (Alexandrov, 1977; Russell, 1990), we have intuitively selected the model of Brandts (1967) to describe the activity of an enzyme which controls the putative growth-rate-limiting reaction of microbial growth.

10.2. THERMODYNAMIC TERMS

This treatment, of necessity, makes use of a number of thermodynamic terms and concepts some of which have been defined previously but which, for ease of comprehension, are presented again here.

10.2.1. Symbols

ΔH: difference in enthalpy, H, (metabolic heat) of two states of a system. In the current discussion, the system may be described as: $X \rightleftharpoons Y$ where X is the catalytically inactive state (D-state) and Y the catalytically active state (N-state) of the growth-rate-limiting enzyme.

ΔS: the difference in entropy of two states of a system. Entropy is essentially a mathematical function and has no simple physical interpretation. It *may* be interpreted as an index of the intrinsic stability of a system, the greater the entropy of a system the smaller is its capacity for spontaneous change. Another interpretation of entropy is that it measures the degree of disorder, or informational content, of a system. Enhanced entropy relates to greater disorder or informational content. Thus, the tendency to increase the net entropy of the system and its surroundings drives spontaneous reactions and the reaction stops when the net entropy of the system has reached its maximum value. At this condition the free energy of the system has fallen to its minimum value.
In a closed system:

$$\Delta S_{system(protein)} + \Delta S_{surroundings(water)}$$

must increase for spontaneous change to occur.

ΔG: the change in Gibbs free energy, under defined but arbitrary conditions of constant temperature and pressure, between two states of the system.

$$\Delta G = \Delta H - T\Delta S \qquad (10.1)$$

For spontaneous reaction to occur in a closed system $\Delta G \leq 0$, where ΔG refers to the difference between the initial and final states. The Gibbs free energy has also often been interpreted as the measure of the ability of a system to perform useful work. An exergonic (energy releasing) reaction is *theoretically* able to perform a maximum amount of useful work equal to ΔG. In this discussion ΔG_{den} represents the amount of work required to disrupt the native cooperative structure of the enzyme. ΔF is an earlier nomenclature for ΔG.

$\Delta G°$: ΔG (ΔF) under the standard conditions: 298.15 K, and standard atmospheric pressure (101325 Pa). Similarly, $\Delta H°$ and $\Delta S°$ are the enthalpy and entropy under the same standard conditions.

C_p: specific heat capacity. The quantity of heat required to raise by one degree the temperature of a unit mass of substance.

ΔC_p: difference in heat capacity between two states of a system; $\Delta C_p = d\Delta H/dT$

K_c: equilibrium constant, calculated as: $K_c = \dfrac{[\text{product}]}{[\text{reactant}]}$

R : (Molar) Gas constant. $R = 8.314\ J\ K^{-1}\ mol^{-1}$

10.2.2. Concepts

ΔG as a function of concentration may be expressed:

$$\Delta G = RT \ln K_c \qquad (10.2)$$

$$\therefore\ \Delta G = RT \ln\left(\frac{[\text{product}]}{[\text{reactant}]}\right)$$

For the situation of the reversible denaturation of protein:

$$\Delta G = RT \ln\left(\frac{[\text{D-state}]}{[\text{N-state}]}\right) \qquad (10.3)$$

ΔG°_{den} (for the change from the N- to the D-state) is small, of the order of 20 - 85 kJ mol^{-1} (Matthews, 1987).

ΔG° has significance because it allows ΔG at other conditions to be estimated from the equation:

$$\Delta G = \Delta G^\circ + RT \ln \frac{\Pi C_p}{\Pi C_r}$$

where ΠC_p and ΠC_r represent the concentration of the reaction 'products' and 'reactants' respectively, each raised to the appropriate power as required by the stoichiometric equation for the process.

For the condition where the concentrations of enzyme in the N- and D-states are equal, Eqn. 10.3 simplifies to:

$$\Delta G^\circ_{den} = RT \ln (1)$$
$$= 0$$

The low and high temperatures, respectively, at which $\Delta G_{den} = 0$ occurs have been variously denoted as follows:

$T_{1/2_L}$, $T_{1/2_H}$	(Schoolfield *et al.*, 1981)
T_L, T_H	(Franks *et al.*, 1990)
$T_G^{den'}$, T_G^{den}	(Privalov and Gill, 1988)

These terms are analogous.

10.3. TEMPERATURE AND ENZYME STABILITY

10.3.1. Enzyme Stability

Privalov and Gill (1988) define the *stability* of enzymes by showing that small globular proteins can exist in one of two dominant conformational states, either native or denatured. The native state is the catalytically active state and is also more compact and retains tertiary and quaternary structure. These authors claim that the compact state of proteins is primarily due to van der Waals interactions between the protein's non-polar groups and that water solvation of these groups leads to destabilisation of the protein. The difference in Gibbs free energy between the N- and D-states is considered to be a measure of the stability.

The folding and unfolding of globular proteins is extremely rapid (*cf*: Richards, 1991) and may be considered to be essentially a two state process (see Chapters 3 and 6), the intermediates being so transient as to be insignificant.

10.3.2. Temperature and Enzyme Stability

Brandts (1964a, b) studied the stability of the folded conformation of globular proteins using the difference spectrum technique. This was based on the fact that Beer's Law was obeyed by both native and denatured protein in the narrow concentration ranges used in his experiments. He summarised the work on the enzymes chymotrypsinogen and ribonuclease (Brandts, 1967) and considered that the five factors shown in Table 10.1 could be summed to give $\Delta G°_{den}$.

Brandts' (1967) representation of the native and denatured states is shown in Fig. 10.1. Brandts (1967) claimed that the N-state conformation is determined by the amino acid sequence and that the conformational state with the lowest Gibbs free energy would be present in the largest amount. The loss of conformation leads to denaturation, and is reversible unless aggregation ensues, i.e.:

$$X \rightleftharpoons Y \rightarrow \text{aggregate}$$

Table 10.1 **Important Structural Features which Contribute to the Difference in Free Energy between Native and Denatured States**

Native		Denatured	Contribution
Rigid structure with little rotational freedom	⇌	Flexible structure	$-pNT\Delta S_c$
Peptide-peptide hydrogen bonds	⇌	Peptide-solvent hydrogen bonds	$pN\Delta h_H$
Non-polar side-chains in protein interior	⇌	Non-polar side-chains exposed to solvent	$p\Sigma\Delta \int_{sc}^{i}$
High charge density due to compactness	⇌	Low charge density in extended conformation	ΔF°_{elec}
Ionisable side-chains with local interactions	⇌	Ionisable side-chains fully solvated	ΔF°_{titr}

Total: $\Delta F^{\circ} = -pNT\Delta S_c + pN\Delta h_H + p\Sigma\Delta \int_{sc}^{i} + \Delta F^{\circ}_{elec} + \Delta F^{\circ}_{titr}$

[*After* Brandts (1967) by permission of Academic Press, Inc., London].

10.3.3. Cold Denaturation

The term *denaturation* refers to structural change in biopolymers which destroys the native, active configuration. In proteins, loss of native structure and biological activity occurs through destruction of tertiary and quaternary structure, with the formation of random coil or metastable forms. In denaturation, non-covalent bonds are broken, in particular, hydrogen bonds. Such denaturation is often reversible. (Scott and Brewer, 1983).

Brandts (1964) predicted that denaturation of protein would occur not only at high temperatures but that loss of activity, due to loss of quaternary and tertiary structure, i.e. denaturation, would also occur at low temperatures. He postulated that this cold denaturation would be due to solvent effects caused by the exposure of hydrophobic groups to the solvent (water). His representation of denaturation (Fig. 10.2) shows that his interpretations of high and low temperature denaturation are quite different. In low temperature denaturation the hydrophobic groups have an ordered solvation shell of water known as a clathrate. In fact, a new compound has been formed which results in the large changes of specific heat (ΔC_p) previously noted in 3.3.4. The main consequence of the change in specific heat capacity is that the thermodynamic functions, ΔH and ΔS, which define the transition between the N- and D-state are both temperature dependent (Privalov and Gill, 1988).

The difference in the heat capacities of the native and denatured proteins of several thousand J $K^{-1} mol^{-1}$ leads to an inversion of the sign of both the low temperature changes in enthalpy and entropy with a maximum stability of native chymotrypsinogen at ~10°C and ribonuclease at ~0°C. This is because the low temperature breakdown of the ordered native structure of proteins leads to a decrease in entropy, i.e. an increase in the order of the solvating water system, which seems paradoxical.

Privalov and Gill (1988) point out that the work of Brandts on cold denaturation of enzymes has not received adequate attention. Only very recently has the Moscow school of Peter Privalov shown, by scanning microcalorimetry on various proteins, that cold denaturation does proceed with the release of heat.

Although the exact nature of the phenomenon is not universally accepted, Franks *et al.* (1990) concluded that "the low and high temperature induced D-states might differ in the sense that at T_L the protein folding is destroyed but secondary (hydrogen-bonded) *domains* [our emphasis] are maintained intact. Such an interpretation is consistent with the remarkable degree of reversibility of cold denaturation, even at higher protein concentrations. On the other hand, the CD and NMR spectra reported for cold- and heat-denatured metmyoglobin exhibit remarkable similarities but differ from those for the N-state, lending support to the existence of a common D-state. ... There can, however, no longer be any doubt about the existence of the phenomenon or its physiological and ecological importance".

10.3.4. Temperature of Maximum Enzyme Stability

Brandts (1967) claimed the maximum stability of an enzyme would occur where $\Delta S°$ (= $-(d\Delta G°/dT)$) = 0. He suggested that the total entropy change arises from two sources, a conformational entropy which will be

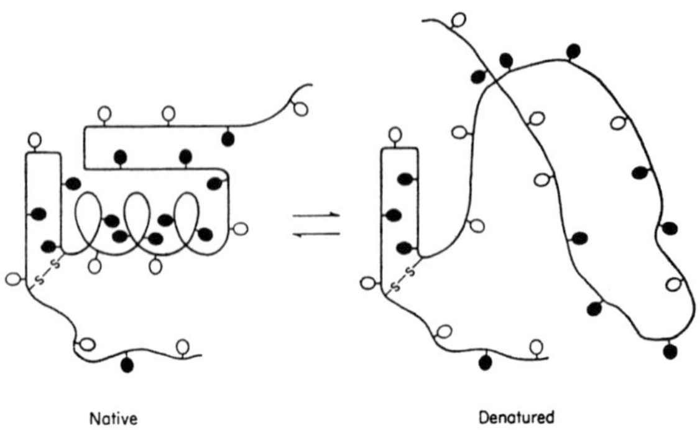

Fig. 10.1 A diagrammatic representation of the denaturation of protein molecules. Hydrophobic side chains (•) are mostly buried in the molecule in the native state, but are exposed to the solvent in the denatured state. Polar and ionic side-chains (o) are largely solvated in both the native and denatured state. [Reproduced from Brandts (1967) with permission from Academic Press, Inc., London].

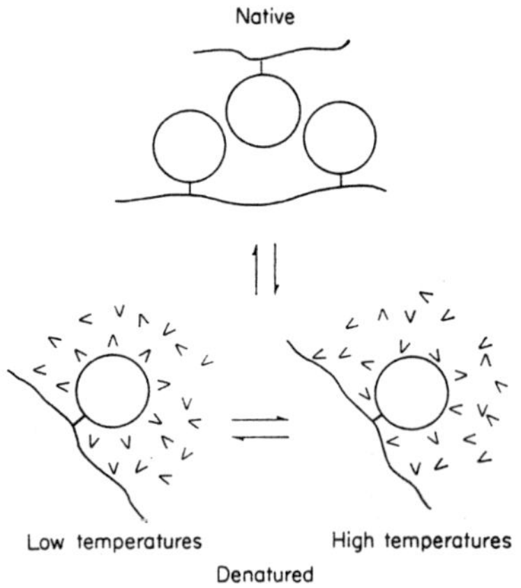

Fig.10.2 A diagrammatic representation of the hydrophobic bonding reaction and the clathrate melting transition. Non-polar side-chains in the native state are surrounded locally by other non-polar groups. In the denatured state, the exposed non-polar groups will be accommodated by an ordered solvation shell of water molecules (represented as v). [Reproduced from Brandts (1967) with permission from Academic Press, Inc., London].

large and positive at all temperatures and an entropy contribution arising from the solvation of hydrophobic groups which will be large and negative at low temperatures. These contributions with opposite signs are equal in value at $\Delta S = 0$.

It should be noted that, in biological systems, the 'T_{max}' of the enzyme is not of necessity the 'T_{opt}' of the reaction. This is because the enzyme-substrate complex will initiate reactions at a higher speed with a true activation energy as temperature increases. The enthalpy and entropy changes in the enzyme caused by raising the temperature above that at which $\Delta S = 0$ eventually outweigh the increase in efficacy of the enzyme-substrate complex.

10.4. APPLICATION OF GIBBS FREE ENERGY TO THE MODELLING OF BACTERIAL GROWTH

10.4.1. Introduction

To incorporate the opposing effects of temperature on the stability of enzymes and the rate of reaction Brandts (1967) developed the model given earlier in 3.3.4. The Gibbs free energy of the enzyme had to be modelled using a power series which would take into account the minimum in $\Delta G°_{den}$ at 'T_{max}'. Brandts (1964b) pointed out that for a great number of amino acids over a wide temperature range the logarithm of the mole fraction solubility in water may be expressed as a power series involving a constant term and three additional terms dependent upon the first, second and third powers of absolute temperature:

$$\Delta G°_{den} = E + FT^1 + GT^2 + HT^3$$

Because he considered that the major changes at low temperature were caused by the large change in free energy due to the solvation of hydrophobic amino acids in the transition zone he substituted :

$$\Delta G°_{trans} = a^i T^1 + b^i T^2 + c^i T^3$$

into Eqn. 3.23. Table 10.2, overleaf, lists values of these constants for eleven amino acids.

10.4.2. Recent Methods for Determining ΔG°_{hyd}

We do not know the putative key enzyme controlling bacterial growth, though Farrell and Rose (1967) surmised that permeases might denature at low temperatures and thus inhibit the transport of monosaccharides, neutral, basic, and acidic amino acids, growth factors and inorganic acids.

The need to know the amino acid composition of the unknown enzyme has been overcome by the work of Murphy *et al.* (1990). These authors have drawn attention to the common features of protein unfolding and the dissolution of hydrophobic groups in general. Pfeil and Privalov (1979) had previously plotted the specific denaturation enthalpies for various enzymes in the high transition range and noted that all data converged to a common value at 112°C (Fig. 10.3, opposite).

It is interesting to note that the highest temperature recorded for growth of hypothermophilic bacteria is 112°C (H. Morgan, *pers. comm.*, 1992). These organisms are recorded to have optimum

Table 10.2 **Classification of the Hydrophobic Side-chains of Amino Acid Residues and the Characteristic Values of the Temperature-independent Constants for these Side-chains (*after* Brandts, 1964b).**

Group	Members	Constants a^i	b^ix 10^2	c^ix10^5
I.	leucine, isoleucine, phenylalanine, proline, tryptophane, tyrosine	-24.906	20.265	-31.330
II.	valine, methionine, lysine (non-polar portion)	-25.746	18.394	-27.354
III.	alanine, arginine, (non-polar portion)	-19.000	13.258	-20.790

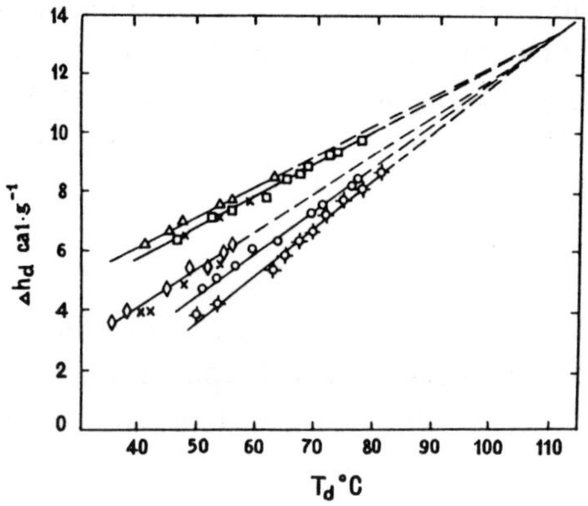

Fig. 10.3 Dependence of specific heat of denaturation on the temperature of transition for seven enzymes. [Reproduced from Pfeil and Privalov (1979) with permission of Elsevier Scientific Publishing Co., Amsterdam].

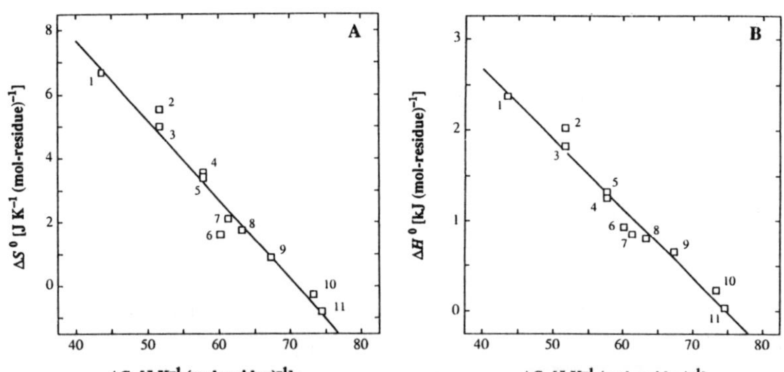

Fig. 10.4 Linear correlation of thermodynamic properties for denaturation at 25°C for the enzymes 1, ribonuclease A; 2, egg white lysozyme; 3, fragment K4 of plasminogen; 4, α-chymotrypsin; 5, β-trypsin; 6, papain; 7, *Staphylococcus* nuclease; 8, carbonic anhydrase; 9, cytochrome *c*; 10, pepsinogen; 11, myoglobin. A: Entropy of denaturation *versus* heat capacity change. B: Enthalpy of denaturation *versus* heat capacity change. [Reproduced from Murphy *et al.* (1990) with permission of Science Publishers].

temperatures for growth of 110°C (Stetter *et al.*, 1989; Huber *et al.*, 1989) although growth temperatures up to 120°C have been reported under higher than atmospheric pressures (Deming and Baross, 1986).

Privalov and Gill (1988) point out that ΔC_p upon denaturation is different for different proteins and seems to correlate with number of contacts between non-polar groups in native proteins. The relation of ΔC_p to $\Delta H°$ and $\Delta S°$ is shown in Fig. 10.4.

10.4.3. Relationship of $\Delta G°$ to $\Delta G°_{hyd}$

Following from Privalov and Gill (1988), Murphy *et al.* (1990) plotted ΔH and ΔS as a function of ΔC_p for groups of compounds at a given temperature. By relating the temperature dependence of ΔH and ΔS to the common temperature of convergence of these functions ($T^*_H = T^*_S \approx 112°C$) they derived the following equations to describe the temperature dependence of these functions for the denaturation of protein:

$$\Delta H = \Delta H^* + \Delta C_p (T - T^*_H) \tag{10.4}$$

$$T\Delta S = T\Delta S^* + \Delta C_p [T \ln (T/T^*_S)] \tag{10.5}$$

The heat capacity terms represent the hydrophobic contributions to the free energy change, namely:

$$\Delta G°_{hyd} = \Delta C_p [(T - T^*_H) - T \ln (T/T^*_S)] \tag{10.6}$$

Murphy *et al.* (1990) stated that the model compound studies suggested that other contributions to the free energy of denaturation are largely insignificant, as had been found by Brandts (1967).

Substituting Eqns. 10.4 and 10.5 into Eqn. 10.1 they obtained:

$$\Delta G = \Delta H^* - T\Delta S^* + \Delta C_p [(T - T^*_H) - T \ln (T/T^*_S)] \tag{10.7}$$

This relationship is represented graphically in Fig. 10.5.

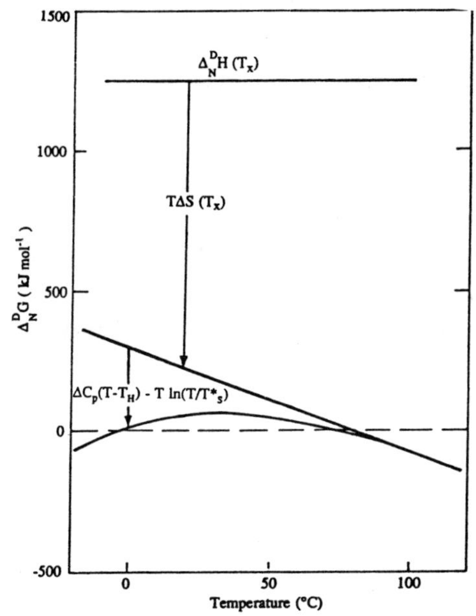

Fig. 10.5 Contribution of dissipative forces [$T\Delta S(T_S)$] and water solvation effect [($\Delta C_P(T - T_H) - T \ln T/T^*_S$)] to the stabilisation of an abstract globular protein consisting of 200 amino acids. [*After* Privalov and Gill (1988)].

10.5. SYNTHESIS OF THE BRANDTS AND THE MURPHY *et al.* MODELS

10.5.1. Construction of a New Thermodynamic Model

The Brandts model, Eqn. 3.24, still does not fulfil the requirement for a fully mechanistic treatment of the thermodynamics of the rate-limiting reaction, because $\Delta G°_{den}$ is modelled by an empirical equation. The Murphy *et al.* (1990) functions, however, are a mechanistic description of ΔG for the thermal denaturation of protein. By substituting Eqn. 10.7 into Eqn. 3.23 we generate the following model to describe the effect of temperature on the rate of enzyme catalysed reactions:

$$\text{rate} = \frac{C\exp(-\Delta E_a / RT)}{1 + \exp\left(-(\Delta H^* - T\,\Delta S^* + \Delta C_p\,[\{T - T^*_H\} - T\ln\{T/T^*_S\}]) / RT\right)}$$

(10.8)

where:

ΔE_a = activation energy of the reaction catalysed by the rate limiting enzyme
ΔH = enthalpy associated with thermal denaturation of the rate limiting enzyme
ΔS = entropy associated with thermal denaturation of the rate limiting enzyme
ΔC_p = difference in heat capacity between the N- and D-state of the rate limiting enzyme
T^*_H = temperature (K) at which the ΔC_p contribution to enthalpy is 0.
T^*_S = temperature (K) at which the ΔC_p contribution to entropy is 0.
$T^*_H = T^*_S \approx 385K$
ΔH^* = value of enthalpy at T^*_H
ΔS^* = value of entropy at T^*_S
T = temperature (K)
R = gas constant (8.314 J K^{-1} mol^{-1})
C = a parameter whose value must be estimated

10.5.2. Predictions from the New Model

We sought to simulate the characteristics of organisms from Table 3.1 representative of each of the four thermal classes of microorganisms. Our early attempts at producing square root plots consistent with those for the selected organisms were unsuccessful, with the exception of *E. coli*, where parameter values used were within the ranges given in Murphy *et al.* (1990). Subsequently, and with the aid of combined spreadsheet and graphing software which enabled the effect of variations in a parameter value to be visualised within a few seconds, we observed that changes in the values of ΔC_p and very small changes in the value of ΔS^* could have profound effects on the predictions. We were able ultimately to produce realistic simulations of rate-temperature plots expected for the organisms chosen.

10.5.3. Relating the New Model to the Square Root Equation

In Fig 10.6 we present the predictions of Eqn. 10.8, for the parameter values given in Table 10.3, as square root plots. For the purposes of comparison we attempted to vary as few parameters as possible for the simulation of the temperature-√growth rate curves. We assumed a rate controlling enzyme of 250 amino acids. ΔH^* was held constant at

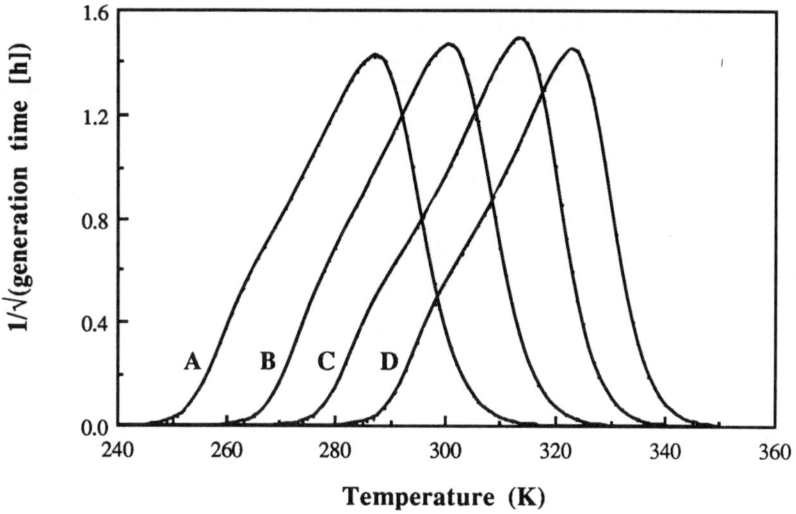

Fig. 10.6 Simulated growth curves for the organisms listed in Table 10.3 based on the predictions of Eqn. 10.9 with the parameter values listed in Table 10.3. A, Antarctic strain 755; B, *Pseudomonas* Group I 16L16; C, *Escherichia coli*; D, *Bacillus* spp. NCIB 12522. Data generated at 1 K intervals, lines through these data obtained by interpolation using CA-Cricket Graph™ 1.3.2[1].

6.4 kJ mol-res^{-1}. Values for ΔC_p and ΔS^* were based initially on values given in Murphy *et al.* (1990), but ΔS^* was subsequently altered empirically by small amounts to produce the desired square root plots. The value for the parameter C was initially assigned to give realistic rate predictions for the *E. coli* simulation. For the other simulations, C was set to produce optimal rate values of the same magnitude as the *E. coli* simulation, to simplify visual comparison of the simulations. ΔE_a values were based on representative published values.

[1] Computer Associates International, Inc., 1240 McKay Drive, San Jose, California, USA.

Table 10.3 **Parameters Values Used with Eqn. 10.8 to Generate Typical Temperature-Growth Curves**

Organism (Thermal Group)	ΔE_a ($kJK^{-1}mol^{-1}$)	C	ΔC_p ($kJK^{-1}mol^{\dagger -1}$)	ΔS^* ($kJK^{-1}mol^{\dagger -1}$)
Antarctic strain (psychrophile)	52.5	$1x10^{10}$	58.0	19.50
Pseudomonas Gp I 16L16 (psychrotroph)	57.5	$3x10^{10}$	66.4	19.05
Escherichia coli (mesophile)	62.5	$2x10^{11}$	75.0	18.68
Bacillus spp. NCIB 12522 (thermophile)	74.0	$2x10^{12}$	89.0	18.30

[$mol^{\dagger -1}$ denotes (mol. amino acid residue^{-1})].

The square root plots initially simulated were for organisms listed in Table 3.1, representative of each of the four recognised thermal classes. Initially, *Bacillus stearothermophilus* 238 was chosen as the thermophilic example, but it was difficult to simulate the cardinal temperatures listed within reasonable ranges for the parameter values. Examination of those cardinal temperatures reveals that they represent unusually narrow temperature limits for growth compared to other organisms in Table 3.1, and *Bacillus* spp. NCIB 12522 was selected instead. Thus our mechanistic model provided a useful framework by which to evaluate predictions from limited observations.

It should be remembered that these plots were, for ease of comparison, generated by varying only four of the parameters of Eqn. 10.8. The simulated curves do not look like 'typical' square root plots; in particular, they do not intercept the temperature axis, but upon examination of Eqn. 10.8 are predicted to achieve zero rate at $\pm \infty$ only. They are more reminiscent of the Schoolfield *et al.* (1981) models fitted by Heitzer *et al.* (1991). In spite of this, and although the temperature-

√growth rate curves are not perfect straight lines in the suboptimal regions, if we now consider the practical aspects of kinetic data collection an explanation for the form of the square root relationship begins to emerge.

Experimenters are constrained in the amount of time that they have available to collect data for long generation times. In our laboratory we would seldom wait for more than four weeks to obtain a growth curve. If we assume that to obtain a growth curve we must follow the growth of the culture from, say, 10^3 to 10^9 CFU/ml, we must wait for approximately 20 generations, discounting the lag phase duration. Over a four week period this suggests that data representative of generation times longer than 34 h would be unlikely to be gathered. Thus, in the square-root plot √rate data values <0.17 would seldom be plotted. The effect of removing these values from our simulated data sets is shown in Fig. 10.7.

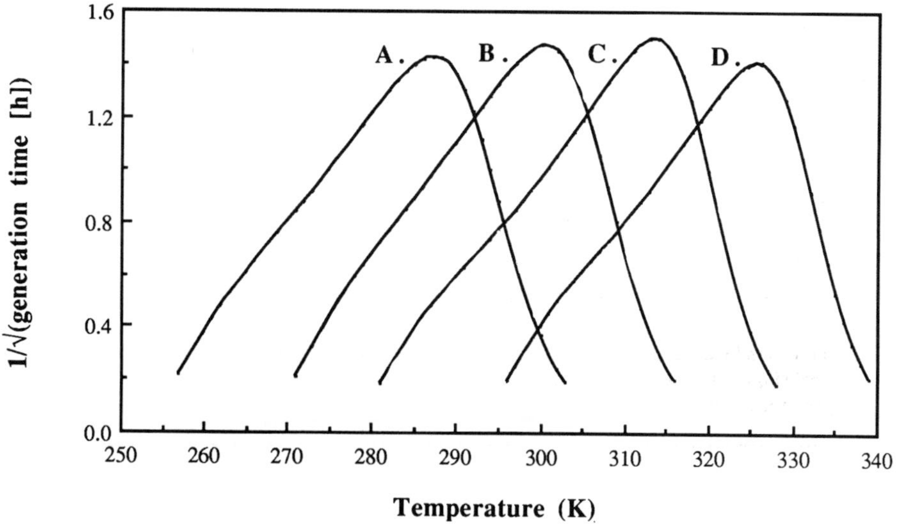

Fig. 10.7 The data of Fig. 10.6 with √rate data values below 0.17 removed to represent the limit of experimental data sets. A, Antarctic strain 755; B, *Pseudomonas* Group I 16L16; C, *E. coli*; D, *Bacillus spp.* NCIB 12522. Data generated at 1 K intervals, lines fitted by interpolation using CA-Cricket Graph™.1.3.2.

This treatment is also consistent with observations that microorganisms have a definite lower temperature limit for growth. In the absence of other environmental stresses this temperature is usually several degrees above their T_{min}.

Simple square-root models (Eqn. 3.10) were then fitted to the suboptimal range of these data sets. The fitted lines are shown in Figs. 10.8 a - d and the results summarised in Table 10.4.

Table 10.4 **Information Summary for Figs. 10.8 a - d**

Organism (Thermal Group)	T_{min} (K)	T_{opt} (K)	T_{max} (K)	R^2
Antarctic strain (psychrophile)	251	287		0.998
	(251)	(288)	(293)	
Pseudomonas Gp I 16L16	265	301		0.997
(psychrotroph)	(266)	(302)	(310)	
E. coli	276	314		0.998
(mesophile)	(276)	(314)	(322)	
Bacillus spp. NCIB 12522	291	324		0.999
(thermophile)	(291)	(324)	(335)	

Numbers in brackets indicate cardinal temperatures estimated from the square root model (Eqn. 3.12) fitted to the original data. In this exercise only the two parameter square root model (Eqn. 3.10) was fitted to the simulated curves.

Figs. 10.8 (Opposite and overleaf) The simulated growth curves of Figs. 10.7 with simple square root models fitted to data in the suboptimal region. Opposite: A, Antarctic strain 755; B, *Pseudomonas* Group I 16L16; Overleaf: C, *Escherichia coli*; D, *Bacillus* spp. NCIB 12522.

A.

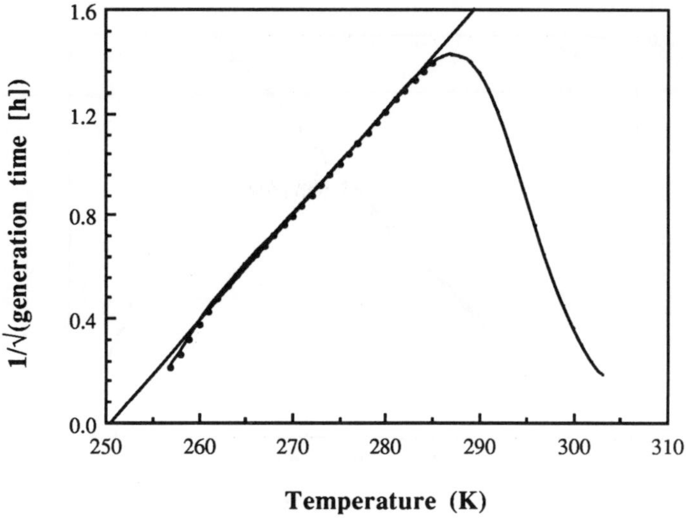
1.6
1.2
0.8
0.4
0.0
1/√(generation time [h])
250
260
270
280
290
300
310
Temperature (K)

B.

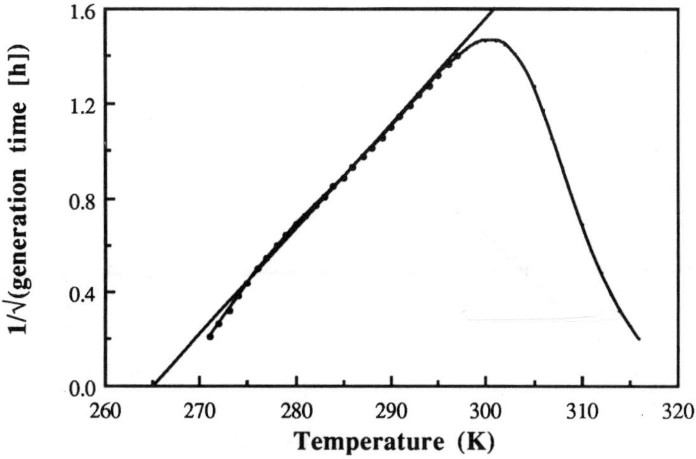
1.6
1.2
0.8
0.4
0.0
1/√(generation time [h])
260
270
280
290
300
310
320
Temperature (K)

C.

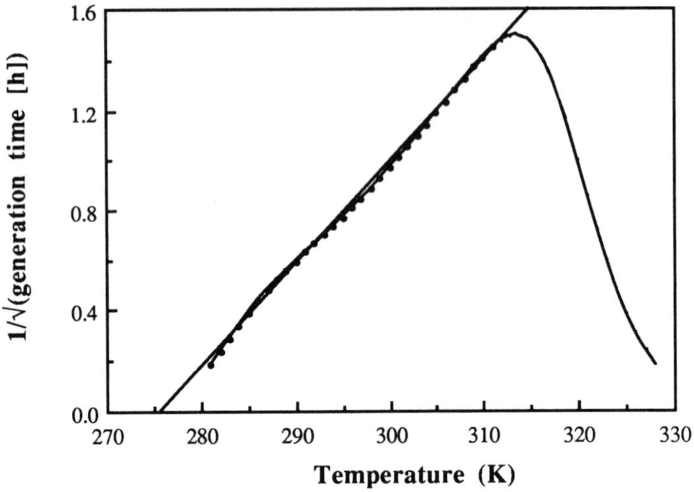
1/√(generation time [h])
1.6
1.2
0.8
0.4
0.0
270
280
290
300
310
320
330
Temperature (K)

D.

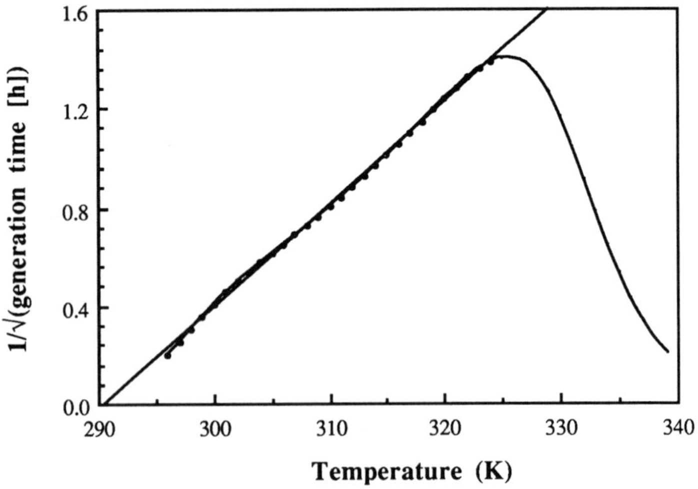
1/√(generation time [h])
1.6
1.2
0.8
0.4
0.0
290
300
310
320
330
340
Temperature (K)

The goodness of the fits, as judged by the R^2 values, is consistent with what is typically achieved using experimental data sets. When one also considers the experimental error and biological variation inherent in the systems used to derive kinetic data, it is evident that the square root model is not inconsistent with the derived mechanistic model, i.e. the two models summarise the same data but in an outwardly different way.

A similar exercise showed that the four parameter square-root model (Eqn. 3.12) describes the data for the full biokinetic range equally well.

10.5.4. Reconciling the Master Reaction Model (Eqn. 10.8) with the Square Root Model

Given good data sets, the ability of both the master reaction-type models and the square root model to describe experimental data well is firmly established in the literature (Zwietering *et al.*, 1991; Heitzer *et al.*, 1991). In this chapter we have shown that data generated by the mechanistic model, Eqn. 10.8, can be described equally well by the square root model, if one considers the limitations of experimental data. Thus we have shown, by example, that there is no inherent contradiction between the two models.

It would, however, be intellectually satisfying to have a more rigorous means of demonstrating this point. Unfortunately, we have been unable to demonstrate any mathematical equivalence between the two model types. Our attempts to do so, however, confirmed the visual evidence that Eqn. 10.8 does not lead to a straight line relationship between temperature and square-root of growth rate in the suboptimal region, but is a near approximation to one.

10.6. PROSPECTS

In this chapter we have developed a mechanistic model consistent with contemporary knowledge of protein thermodynamics. This model is able to simulate well a wide range of bacterial temperature-growth rate curves, using realistic values for the parameters of the equation. The model could, perhaps, be further improved by the inclusion of a temperature term in the numerator to model the known temperature dependence of the 'collision factor', related to the parameter C in Eqn. 10.8. But do we really need another 'mechanistic' model? We have already pointed out that the model is incomplete in that it fails to take into account irreversible enzyme denaturation. Furthermore, we do not know the identity of the putative key enzyme. Heitzer *et al.* (1991) pointed out that a 'mechanistic' model whose parameters cannot be determined experimentally cannot be considered truly to be mechanistic.

Nonetheless, Eqn. 10.8 represents a hypothesis, and as such it has value if it helps to explain observations and provide a framework for increasing understanding and posing appropriate research questions. On a purely mathematical level, we have shown that Eqn. 10.8 can be used as a 'mechanistic' model of four parameters which models the entire biokinetic range. Privalov and Gill (1988), on the basis of the concepts presented here, have shown that T_G^{den} and $T_G^{den'}$ are not independent but related by the following equation:

$$T_G^{den'} \approx \frac{\left(T_G^{den}\right)^2}{3\,T_G^{den} - 2\,T_H^{den}}$$

where T_H^{den} is the temperature at which the enthalpy of denaturation is zero. This provides an explanation, for example, for previous observations (Lowry and Ratkowsky, 1983; Ratkowsky, 1990) that the six parameter Schoolfield model (Eqn. 3.19) is over-parameterised. The difficulty experienced in fitting Eqn. 10.8 to cardinal temperatures derived from a square-root model for the *B. stearothermophilus* 238 data set, led us to reevaluate the data set and subsequently recognise that it was anomalous. Perhaps of greater interest is that an analysis of Eqn. 10.8 has led us to a rational explanation for the form of the square root model. It also reinforces the status of T_{min} as a notional temperature of no metabolic significance. From the mechanistic viewpoint developed in this chapter T_{min} is shown to be simply the intercept of the extrapolation of the linear (suboptimal) portion of a square root of growth rate plot with the temperature axis.

The fact that the Bělehrádek power function applies to organisms from psychrophiles through to thermophiles, and for both the eu- and the archaebacteria (McMeekin and Franzmann, 1988), might indicate that we are dealing with a universal phenomenon such as the unfolding of hydrophobic side-chains of key proteins. This universality is reflected in the ability of Eqn. 10.8 to generate realistic simulations of growth curves, whilst adhering to narrow limits of parameter values. The sensitivity of the curves generated to seemingly small changes in parameter values is also consistent with recent findings of protein stability. Matthews' (1987) review indicates that the substitution of a few amino acids in the primary structure of the protein can cause dramatic changes in stability without significantly altering the protein's tertiary structure. The effect of such substitutions is usually to cause small, localised changes in tertiary structure leading to movements of up to 1 Å in the position of side-chain atoms and 0.5 Å or less in the position of backbone atoms.

Gounot (1991) points out that both ribosomal proteins and regulatory enzymes require a precise conformation to be functional. She supports the theory developed in this chapter and also espoused by Jaenicke (1990) and Dill (1990), that the weakening of hydrophobic bonds at low temperature results in conformational changes. Genes encode for the amino acid composition of regulatory proteins. She arranges the hydrophobic residues from thermophilic to psychrophilic: Phe, Val, Ile, Leu, Ala, Met (see Table 10.2). Schlatter *et al.* (1987) and Zuber (1988) have found temperature related amino acid substitutions in the lactic dehydrogenases from thermophilic, mesophilic, and psychrotrophic *Bacillus* spp.

We have not complied strictly to the results of Murphy *et al.* (1990) for the constancy of ΔH* and ΔS*, but have allowed ΔS* to alter slightly to achieve the simulated temperature-√growth rate plots presented. However, the proteins used by Murphy *et al.* (1990) were all from homeothermic organisms, mostly mammals. It is possible that different values for these parameters would be obtained for proteins from organisms adapted to different thermal environments.

Thus we begin to see how thermodynamic parameters can be manipulated to give the simulation of typical square root plots throughout the biokinetic temperature range. We are thus foreshadowing Cole's (1991) defence of predictive microbiology when he stated: "Researchers in the field of predictive modelling are striving to develop models for microbial growth and death based upon an understanding of cell variability and physiology and that could be used to extrapolate to other conditions".

REFERENCES

Adair, C., Kilsby, D.C. and Whittall, P.T. (1989). Comparison of the Schoolfield (non-linear Arrhenius) model and the Square Root model for predicting bacterial growth in foods. *Food Microbiology*, **6**: 7-18.

Adair, C., Briggs, P.A., Cleaver,G.J. and Schofield, G.M. (*in press*). The concept and application of expert systems in the field of microbiological safety. *Proceedings of the International Workshop on the Application of Predictive Microbiology and Computer Modeling Techniques to the Food Industry, Tampa, Florida, April 1992.* Sponsored by Society for Industrial Microbiology.

Adams, M.R., Little, C.L. and Easter, M.C. (1991). Modelling the effect of pH, acidulant and temperature on the growth rate of *Yersinia enterocolitica. Journal of Applied Bacteriology*, **71**: 65-71.

Alexandrov, V. Ya. (1977). *Cells, Molecules and Temperature. Conformational Flexibility of Macromolecules and Ecological Adaptation.* (Translated from the Russian by V. A. Bernstam.) Springer-Verlag, Berlin.

Anon. (1990). *Predictive modelling of microorganisms in foods: protocols documents for the production and recording of data.* Campden Food and Drink Research Association, Chipping Campden, U.K.

Arrhenius, S. (1889). Über die Reaktiongeschwindigkeit bei der Inversion von Rohrzucker durch Säuren. *Zeitschrift für Physiological Chemie*, **4**: 226 - 248.

Arrhenius, S. (1915). *Quantitative Laws in Biological Chemistry.* G. Bell and Sons Limited, London. Reproduced, Readex Microprints, New York, 1973.

Baird-Parker, A.C. and Kilsby, D.C. (1987). Principles of predictive food microbiology. *Journal of Applied Bacteriology*, Symposium Supplement pp. 43S-49S.

Barnes, E.M., Impey, C.S., Geeson, J.D. and Buhagiar, R.W.M. (1978). The effect of storage temperature on the shelf life of eviscerated air-chilled turkeys. *British Poultry Science*, **19**: 77-84.

Barlow, S.M. (1976). Fish meal manufacture in the tropics. In *Handling, Processing and Marketing of Tropical Fish.* Tropical Products Institute, London pp. 223-229.

Bates, D.M. and Watts, D.G. (1980). Relative curvature measures of nonlinearity. *Journal of the Royal Statistical Society, Series B.*, **42**: 1-25.

Bates, D.M. and Watts, D.G. (1988). *Nonlinear Regression Analysis and its Applications*. John Wiley and Sons, New York.

Baskerville, G.L. (1972). Use of logarithmic regression in the estimation of plant biomass. *Canadian Journal of Forestry*, **2**: 49-53.

Bělehrádek, J. (1926a). Influence of temperature on biological processes. *Nature (London)*, **118**: 117-118.

Bělehrádek, J. (1926b). Protoplasmic viscosity as determined by a temperature coefficient of biological reactions. *Nature (London)*, **118:** 478-480.

Bělehrádek, J. (1930). Temperature coefficients in biology. *Biological Reviews and Biological Proceedings of the Cambridge Philosophical Society*, **5**: 30-58.

Bělehrádek, J. (1935). *Temperature and Living Matter*. Gebrüder Borntraeger, Berlin.

Bělehrádek, J. (1954). Temperature and rate of enzyme action. *Nature (London)*, **173**: 70-71.

Bělehrádek, J. (1959). Physiological aspects of heat and cold. *Annual Review of Physiology*, **19**: 59-81.

Belsley, D.A., Kuh, E. and Welsch, R.E. (1980). *Regression Diagnostics: Identifying Influential Data and Sources of Variation*. John Wiley and Sons, New York.

Beuchat, L.R. (1987). Influence of water activity on sporulation, germination, outgrowth and toxin production. In *Water Activity: Theory and Applications to Foods*. L.B. Rockland and L.R. Beuchat *Eds*. Marcel Dekker Inc., New York, pp. 139-151.

Blankenship, L.C., Craven, S.E., Leffler, R.G. and Custer, C. (1988). Growth of *Clostridium perfringens* in cooked chili during cooling. *Applied and Environmental Microbiology*, **54**: 1104-1108.

Blixt, K.G. (1983). The I-Point TTM - a versatile biochemical time-temperature integrator. *IIR-XVIth International Congress of Refrigeration. Comm. C.2. Proceedings*, **3**: 793-796.

Blixt, K.G. and Tiru, M. (1977). An enzymatic time/temperature device for monitoring the handling of perishable commodities. *Developmental Biology Standard*, **36**: 237-241.

Boyd, R.F. (1988). *General Microbiology*, 2nd Edition. Times Mirron/Mosby College Publishing, St. Louis.

Brandts, J.F. (1964a). The thermodynamics of protein denaturation. I. The denaturation of chymotrypsinogen. *Journal of the American Chemical Society*, **86**: 4291-4301.

Brandts, J.F. (1964b). The thermodynamics of protein denaturation. II. A model of reversible denaturation and interpretations regarding the stability of chymotrypsinogen. *Journal of the American Chemical Society*, **86**: 4302-4314.

Brandts, J.F. (1967). Heat effects on proteins and enzymes. Chapter 3 in *Thermobiology*, A.H. Rose, *Ed.* Academic Press, New York., pp. 25-72.

Brandts, J.F. (1969). Conformational transitions of proteins in water and in aqueous mixtures. In *Structure and Stability of Biological Macromolecules*, S.N. Timasheff and G.D. Fasman *Eds.* Marcel Dekker, New York, pp. 213-290.

Bratchell, N., Gibson, A.M., Truman, M., Kelly, T.M. and Roberts, T.A. (1989). Predicting microbial growth: the consequences of quantity of data. *International Journal of Food Microbiology*, **8**: 47-58.

Bratchell, N., McLure, P.J., Kelly, T.M. and Roberts, T.A. (1990). Predicting microbial growth: graphical methods for comparing models. *International Journal of Food Microbiology*, **11**: 279-288.

Bremner, H.A. (1985). A convenient, easy to use system for estimating the quality of chilled seafoods. *DSIR Fish Processing Bulletin*, No. 7. Fish Processing Conference '85, Nelson, New Zealand. Department of Science and Industrial Research, Wellington, New Zealand, pp. 59-70.

Bremner, H.A., Olley, J. and Vail, A.M.A. (1987). Estimating time-temperature effects by a rapid systematic sensory method. In *Seafood Quality Determination*. D.E. Kramer and J. Lister, *Eds.* Elsevier Science Publishers, Amsterdam, pp. 413-435.

Brignell, J.E. and Dorey, A-P. (1983). Sensors for microprocessor-based applications. *Journal of Physics. E. Scientific Instruments*, **16**: 952-958.

Broughall, J.M. and Brown, C. (1984). Hazard analysis applied to microbial growth in foods: development and application of three-dimensional models to predict bacterial growth. *Food Microbiology*, **1**: 13-22.

Broughall, J.M., Anslow, P. and Kilsby, D.C. (1983). Hazard analysis applied to microbial growth in foods: development of mathematical models describing the effect of water activity. *Journal of Applied Bacteriology*, **55**: 101-110.

Brownlie, L.E. (1969). *Effect of some environmental factors on the growth of* Microbacterium thermosphactum *with particular reference to the storage of meat.* M.Sc.Thesis, University of Sydney, Australia.

Buchanan, R.L. (1991). Using spreadsheet software for predictive microbiology applications. *Journal of Food Safety*, **11**: 123-124.

Buchanan, R.L., Stahl, H.G. and Whiting, R.C. (1989). Effects and interactions of temperature, pH, atmosphere, sodium chloride and sodium nitrite on the growth of *Listeria monocytogenes*. *Journal of Food Protection*, **52**: 844-851.

Buchanan, R.L. and Cygnarowicz, M.L. (1990). A mathematical approach to defining and calculating the duration of the lag phase. *Food Microbiology*, **7**: 237-240.

Buchanan, R.L. and Klawitter, L.A. (1991). Effect of temperature history on the growth of *Listeria monocytogenes* Scott A at refrigeration temperatures. *International Journal of Food Microbiology*, **12**: 235-246.

Buchanan, R.L. and Phillips, J.G. (1990). Response surface model for predicting the effects of temperature, pH, sodium chloride content, sodium nitrite concentration and atmosphere on the growth of *Listeria monocytogenes*. *Journal of Food Protection*, **53**: 370-376, 381.

Campbell-Platt, G. (1985). Microbiological spoilage of chilled foods. *Proceedings of the Institute of Food Science and Technology, United Kingdom*, **18**: 2-6.

Carseldine, A.J. and Weste, R.R. (1976). Time-temperature history recorders. *CSIRO Food Research Quarterly*, **36**: 41-45.

Causton, D.R. (1977). *A Biologist's Mathematics*. Arnold, London.

Causton, D.R. (1987). *A Biologist's Advanced Mathematics*. Allen and Unwin, London.

Chandler, R.E. (1988). *The effect of temperature and water activity on microbial growth rate and food spoilage*. PhD Thesis, University of Tasmania.

Chandler, R.E. and McMeekin, T.A. (1985a). Temperature function integration and the prediction of the shelf-life of milk. *Australian Journal of Dairy Technology*, **40**: 10-13.

Chandler, R.E. and McMeekin, T.A. (1985b). Temperature function integration and its relationship to the spoilage of pasteurized homogenized milk. *Australian Journal of Dairy Technology*, **40**: 37-41.

Chandler, R.E. and McMeekin, T.A. (1989a). Modelling the growth response of *Staphylococcus xylosus* to changes in temperature and glycerol concentration/water activity. *Journal of Applied Bacteriology*, **66**: 543-548.

Chandler, R.E. and McMeekin, T.A. (1989b). Combined effect of temperature and salt concentration/water activity on the growth rate of *Halobacterium* spp. *Journal of Applied Bacteriology*, **67**: 71-76.

Chandler, R.E. and McMeekin, T.A. (1989c). Temperature function integration as the basis of an accelerated method to predict the shelf life of pasteurized-homogenized milk. *Food Microbiology*, **6**: 105-111.

Charm, S.E., Learson, R.J., Ronsivalli, L.J. and Schwartz, M. (1972). Organoleptic technique predicts refrigeration shelf life of fish. *Food Technology*, **26**: 65-68.

Charnock, J.S. (1978). Membrane lipid phase-transitions: a possible biological response to hibernation? In *Natural Torpidity and Thermogenesis*. L.C.H. Wang and J.W. Hudson *Eds*. Academic Press, New York. pp. 417-460.

Cochran, W.G. and Cox, G.M. (1957). *Experimental designs.* 2nd Edition.. Wiley, New York.

Cold Chain Newsletter (1989). Competitor for 3M emerges. *Expanded Programme on Immunization.* World Health Organization, Geneva, **89.3**: 5.

Cole, M.B. (1991). *Letters in Applied Microbiology,* **13**: 218-219.

Cole, M.B. and Keenan, M.H.J. (1987). A quantitative method for estimating shelf life of soft drinks using a model system. *Journal of Industrial Microbiology*, **2**: 59-62.

Collins-Thompson, D.L., Wood, D.S. and Beveridge, T.J. (1983). Characteristics of the inhibition of *Brocothrix thermosphacta* by *Lactobacillus brevis. Journal of Food Protection*, **46**: 403-407.

Connell, J.J. (1975). *Fish Quality.* Fishing News (Books) Limited, West Byfleet, England, p. 177.

Cook, R.D. and Weisberg, S. (1982). *Residuals and Influence in Regression.* Chapman and Hall, London.

Daud, H.B., McMeekin, T.A. and Olley, J. (1978). Temperature function integration and the development and metabolism of poultry spoilage bacteria. *Applied and Environmental Microbiology*, **36**: 650-654.

Davey, K.R. (1989a). Comparison of the Schoolfield (non-linear Arrhenius) model and the Square Root model for predicting bacterial growth in foods - a reply to C. Adair *et al. Food Microbiology,* **6**: 302-303.

Davey, K.R. (1989b). A predictive model for combined temperature and water activity on microbial growth during the growth phase. *Journal of Applied Bacteriology,* **67**: 483-488.

Davey, K.R. (1991). Applicability of the Davey (linear Arrhenius) predictive model to the lag phase of microbial growth. *Journal of Applied Bacteriology*, **70**: 253-257.

Davey, K.R., Lin, S.H. and Wood, D.G. (1978). The effect of pH on continuous high temperature/short time sterilisation of liquid. *American Institute of Chemical Engineers Journal*, **24**: 537-540.

Dawes, E.A. (1964). Enzyme kinetics (optimum pH, temperature and activation energy). In *Comprehensive Biochemistry.* M. Florkin and E. H. Stotz, *Eds.* Elsevier Publishing Co., Amsterdam, pp. 89-125.

Deming, J.W. and Baross, J.A. (1986). Solid medium for culturing black smoker bacteria at temperatures to 120°C. *Applied and Environmental Microbiology*, **51**: 238-243.

Department of Industry, Technology and Commerce, Australia (1989). *Food Processing*, p. 24.

Dill, K.A. (1990). Dominant forces in protein folding. *Biochemistry*, **29**: 7133-7155.

Doe, P.E. and Heruwati, E. (1988). A model for the prediction of the microbial spoilage of sun-dried tropical fish. *Journal of Food Engineering*, **8**: 47-72.

Draper, N.R. (1988). Response surface designs. In *Encyclopaedia of Statistical Sciences*. S. Kotz and N.L. Johnson *Eds*. Wiley-Interscience, John Wiley and Sons, New York, **8**: 107-119.

Draper, N.R. and Smith, H. (1981). *Applied Regression Analysis*, 2nd Edition. John Wiley and Sons, New York.

Elliot, R.P. and Michener, H.D. (1965). Factors affecting the growth of psychrophilic microorganisms in foods. *US Department of Agriculture, Technical Bulletin*, **320**: 1-28.

Erickson, R.O. (1970). *Encyclopedia of the Biological Sciences,* 2nd Edition. P. Gray Van Nostrand Reinhold Company, New York, p. 388.

Eyring, H. (1935). The activated complex in chemical reactions. *Journal of Chemical Physics*, **3**: 107-115.

Farber, J.M. (1986). Predictive modelling of food deterioration and safety. In *Foodborne Microorganisms and their Toxins*, M.D. Pierson and N.J. Stern *Eds*. Marcel Dekker Inc. pp. 57-90.

Farrell, J. and Rose, A.H. (1967). Temperature effects on microorganisms. Chapter 6 in *Thermobiology*. A.H. Rose *Ed*. Academic Press, New York, pp.147-218.

Fields, S.C. and Prusik, T. (1986). Shelf life estimation of beverage and food products using bar coded time-temperature indicator labels. In *The Shelf Life of Foods and Beverages*. G. Charalambous *Ed*. Elsevier Science Publishers BV, Amsterdam, pp. 85-95.

Finney, D.J. (1941). On the distribution of a variate whose logarithm is normally distributed. *Journal of the Royal Statistical Society, Series B*, **7**: 155-161

Fiolitakis, E., Grobbelaar, U., Hegewald, E. and Soeder, C.J. (1986). Deterministic interpretation of the temperature response of bacterial growth. *Biotechnology and Bioengineering*, **30**: 541-547.

Firstenberg-Eden, R. and Eden, G. (1984). *Impedance Microbiology*. Innovation in Microbiology Series, Vol. 3, A.N. Sharpe *Series Ed*. Research Studies Press Limited, Letchworth, Hertfordshire, England.

Folks, J.L. (1983). Inverse Gaussian distribution. Volume 4 in *Encyclopaedia of Statistical Sciences*, S. Kotz and N.L. Johnson, *Eds*. John Wiley and Sons, New York, pp. 246-249

Food Control AB. (1974). Improvements in and relating to methods and devices for indicating temperature conditions with respect to time. *Patent Specification 1366797* Patent Office, London.

Franks, F., Mathias, S.F., and Hatley, R.H.M. (1990). Water, temperature and life. *Philosophical Transactions of the Royal Society, London. B,* **326**: 517-533.

Friedman, J.H. (1984). A variable span smoother. *Technical Report No. 5, Laboratory for Computational Statistics*, Stanford University, California, USA.

Fu, B., Taoukis, P.S. and Labuza, T.P. (1991). Predictive microbiology for monitoring spoilage of dairy products with time-temperature integrators. *Journal of Food Science*, **56**: 1209-1215.

Garthright, W.E. (1991). Refinements in the prediction of microbial growth curves. *Food Microbiology*, **8**: 239-248.

Gelman, A., Pasteur, R. and Rave, M. (1990). Quality changes and storage life of common carp (*Cyprinus carpio*) at various storage temperatures. *Journal of the Science of Food and Agriculture*, **52**: 231-247.

GENSTAT (1987). *Genstat 5 Reference Manual.* Oxford, Clarendon Press.

Gibbs, P.A. and Williams, A.P. (1990). Using mathematics for shelf life prediction. *Food Technology International, Europe.* A. Turner Sterling Publications International Limited, London, pp. 287-290.

Gibson, D.M. (1985). Predicting the shelf life of packaged fish from conductance measurements. *Journal of Applied Bacteriology*, **58**: 465-470.

Gibson, A.M., and Roberts, T.A. (1986a). The effect of pH, water activity, sodium nitrite and storage temperature on the growth of *Escherichia coli* and salmonellae in a laboratory medium. *International Journal of Food Microbiology*, **3**: 183-194.

Gibson, A.M., and Roberts, T.A. (1986b). The effect of pH, sodium chloride, sodium nitrite and storage temperature on the growth of *Clostridium perfringens* and faecal streptococci in a laboratory medium. *International Journal of Food Microbiology*, **3**: 195-210.

Gibson, A.M., Bratchell, N. and Roberts, T.A. (1987). The effect of sodium chloride and temperature on the rate and extent of growth of *Clostridium botulinum* type A in pasteurized pork slurry. *Journal of Applied Bacteriology*, **62**: 479-490.

Gibson, A.M., Bratchell, N. and Roberts, T.A. (1988). Predicting microbial growth: growth responses of salmonellae in a laboratory medium as affected by pH, sodium chloride and storage temperature. *International Journal of Food Microbiology*, **6**: 155-178.

Gildberg, A. (1982). *Autolysis of fish tissues: General aspects.* DSc Thesis, Institute of Fisheries, University of Tromsø.

Gill, C.O. (1984). Prevention of early spoilage of livers. *Proceedings of the 30th European Meeting of Meat Research Workers, Bristol, U.K.*, pp. 240-241.

Gill, C.O. (1986). Temperature function integration for hygiene evaluation of food processing procedures. *Food Technology in Australia*, **38**: 203-204.

Gill, C.O. and Harrison, J.C.L. (1985). Evaluation of the hygienic efficiency of offal cooling procedures. *Food Microbiology*, **2**: 63-69.

Gill, C.O. and Newton, K.G. (1977). The development of the aerobic spoilage flora on meat stored at chill temperatures. *Journal of Applied Bacteriology*, **43**: 189-195.

Gill, G.O. and Newton, K.G. (1980). Growth of bacteria on meat at room temperature. *Journal of Applied Bacteriology*, **49**: 315-323.

Gill, C.O., Phillips, D.M. and Loeffen, M.P.F. (1988a). A computer program for assessing the remaining storage life of chilled red meats from product temperature histories. In *Refrigeration for Food and People: Proceedings of Meetings of Commissions C2, D1, D2/3, E1 (September 5 - 9, 1988)*, Institut International du Froid - International Institute of Refrigeration, Paris, pp. 73-77.

Gill, C.O., Phillips, D.M., Loeffen, M.P.F. and Bishop, C. (1988b). A computer program for evaluating the hygienic efficiency of meat processing procedures from product temperature history data. *Congress Proceedings: 34th International Congress of Meat Science and Technology, Brisbane, Australia.* pp. 531-532.

Gill, C.O., Harrison, J.C.L. and Phillips, D.M. (1991a). Use of a temperature function integration technique to assess the hygienic adequacy of a beef carcass cooling process. *Food Microbiology*, **8**: 83-94.

Gill, C.O., Jones, S.D.M. and Tong, A.K.W. (1991b). Application of a temperature function integration technique to assess the hygienic adequacy of a process for spray chilling beef carcasses. *Journal of Food Protection*, **54**: 731-736.

Glasstone, S. (1960). *Textbook of Physical Chemistry*, 2nd Edition. Macmillan and Company Limited, London, pp. 1088-1091.

GLIM (Generalised Linear Interactive Modelling) (1978). Numerical Algorithms Group, Oxford, U.K.

Gompertz, B. (1825). On the nature of the function expressiveness of the law of human mortality, and on a new mode of determining the value of life contingencies. *Philosophical Transactions of the Royal Society, London*, **115**: 513-585.

Gould, G.G. (1989). Predictive mathematical modelling of microbial growth and survival in foods. *Food Science and Technology Today*, **3(2)**: 89-92.

Gounot, A.M. (1991). Bacterial life at low temperature: physiological aspects and biotechnological implications. *Journal of Applied Bacteriology*, **71**: 386-397.

Griffiths, M.W. and Phillips, J.D. (1988a). Modelling the relation between bacterial growth and storage temperature in pasteurized milks of varying hygienic quality. *Journal of the Society of Dairy Technology*, **41**: 96-102.

Griffiths, M.W. and Phillips, J.D. (1988b). Prediction of the shelf life of pasteurised milk at different temperatures. *Journal of Applied Bacteriology*, **65**: 269-278.

Griffiths, M.W., Phillips, J.D. and Muir, D.D. (1987). Effect of low temperature storage on the bacteriological quality of raw milk. *Food Microbiology*, **4**: 285-291.

Grisius, R., Wells, J.H., Barrett, E.L. and Singh, R.P. (1987). Correlation of time-temperature indicator response with microbial growth in pasteurized milk. *Journal of Food Processing and Preservation*, **11**: 309-324.

Gurney, R.W. (1953). *Ionic Processes in Solution*. McGraw Hill Book Company Inc., New York.

Hayward, L.J. (1990). *Predictive microbiology of* Aeromonas hydrophila. *The effect of temperature and water activity on the growth rate of* Aeromonas hydrophila. BSc(Hons) Thesis, University of Tasmania.

Heitzer, A., Kohler, H.E., Reichert, P. and Hamer, G. (1991). Utility of phenomenological models for describing temperature dependence of bacterial growth. *Applied and Environmental Microbiology*, **57**: 2656-2665.

Huber, R., Kurr, M., Jannasch, H. W. and Stetter, K.O. (1989). A novel group of abyssal methogenic archaebacteria (*Methanopyrus*) growing at 110°C. *Nature (London)*, **342**: 833-834.

Hultin, E. (1955). The influence of temperature on the rate of enzymic processes. *Acta Chemica Scandinavica*, **9**: 1700-1710.

ICMSF (1988). *Microorganisms in foods. 4. Application of the hazard analysis critical control point (HACCP) system to ensure microbiological safety and quality*. Blackwell Scientific Publications, Oxford.

Ingraham, J.L., Maaløe, O. and Neidhardt, F.C. (1983). Growth of cells and cultures. Chapter 5 in *Growth of the Bacterial Cell*. Sinauer Associates, Sunderland, Mass., pp. 227-265.

I-Point®. *Product Manual for the Protection of Your Investment in Quality*. I-Point Technologies AB, Nya Agnesfridsvagen 181, S-21375, Malmö, Sweden. (No longer in operation.)

IUMS-ICFMH (1991). A Workshop - Behaviour of Gram-negative pathogens on refrigerated foods. *International Journal of Food Microbiology*, **13**: 183-238.

Jaenicke, R. (1990). Structure and function at low temperatures. *Proceedings and Transactions of the Royal Society, London*, **B326**: 535-553.

James, S. and Evans, J. (1990). Performance of domestic refrigerators and retail display cabinets for chilled products. *Progress in the Science and Technology of Refrigeration in Food Engineering*, Commissions B2, C2, D1, D2/3, Dresden, Germany. International Institute of Refrigeration, Paris, pp. 401-409.

Jarvis, B. (1989). *Statistical Aspects of the Microbiological Analysis of Foods*. Elsevier, Amsterdam.

Jason, A.C. (1983). A deterministic model for monophasic growth of batch cultures of bacteria. *Antonie van Leeuwenhoek*, **49**: 523-536.

Jason, A.C. and Richards, J.C.S. (1975). The development of an electronic fish freshness meter. *Journal of Physics. E: Scientific Instruments*, **8**: 826-830.

Jefferies, C.J. and Brain, P. (1984). A mathematical model of pollen-tube penetration in apple styles. *Planta*, **160**: 52-58.

Johnson, F.H. and Lewin, I. (1946). The growth rate of *E. coli* in relation to temperature, quinine and coenzyme. *Journal of Cellular and Comparative Physiology*, **28**: 47-75.

Johnson, F.H., Eyring, H. and Polissar, M.J. (1954). Temperature. Chapter 8 in *The Kinetic Basis of Molecular Biology*, John Wiley and Sons Inc., New York., pp. 187-285.

Johnson, F.H., Eyring, H. and Stover, B.J. (1974). *The Theory of Rate Processes in Biology and Medicine*. John Wiley and Sons, New York, pp. 199.

Jones, J.E. (*in press*). A real-time database/models base/expert system in predictive microbiology. *Proceedings of the International Workshop on the Application of Predictive Microbiology and Computer Modeling Techniques to the Food Industry, Tampa, Florida, April 1992*. Sponsored by Society for Industrial Microbiology.

Jones, J.E. and Walker, S.J. (*in press*). Advances in modelling microbial growth. *Proceedings of the International Workshop on the Application of Predictive Microbiology and Computer Modeling Techniques to the Food Industry, Tampa, Florida, April 1992*. Sponsored by Society for Industrial Microbiology.

Kilsby, D.C. (1989). Comparison of the Schoolfield (non-linear Arrhenius) model and the Square Root model for predicting bacterial growth in foods - a reply to K.R. Davey. *Food Microbiology*, **6**: 308-309.

Kuiper, F. and Kuiper, P.J.C. (1978). Application of the entropy-enthalpy relationship in water transport through plant roots. *Physiologica Plantarum*, **43**: 121-125.

Labuza, T.P. and Taoukis, P.S. (1991). The relationship between processing and shelf-life. In *Food for the 90's*. Gordon Birch *Ed*. Plenum Press, New York, pp. 73-106.

Langeveld, L.P.M. (1983). Bacteria and the keeping quality of pasteurized milk. *Netherlands Milk Dairy Journal*, **37**: 244-245.

Langeveld, L.P.M. and Cuperus, F. (1980). The relation between temperature and growth rate in pasteurized milk of different types of bacteria which are important to the deterioration of that milk. *Netherlands Milk Dairy Journal*, **34**: 106-125.

Lanigan, G.W. (1963). Silage bacteriology. 1. Water activity and temperature relationships of *Lactobacillus plantarum*, *Lactobacillus brevis* and *Pediococcus cerevisiae*. *Australian Journal of Biological Sciences*, **16**: 606-615.

Leistner, L. (1987). Shelf-stable products and intermediate moisture foods based on meat. In *Water Activity: Theory and Applications to Food.* L.B. Rockland and L.R. Beuchat *Eds.* Marcel Dekker Inc., New York, pp. 295-327.

Leistner, L. and Rodel, W. (1976). Inhibition of microorganisms in food by water activity. In *Inhibition and Inactivation of Vegetative Microbes.* F.A. Skinner and W.B. Hugo *Eds.* Academic Press, London, pp. 219-237.

Li, W.K.W. and Dickie, P.M. (1987). Temperature characteristics of photosynthetic and heterotrophic activities: seasonal variations in temperate microbial plankton. *Applied and Environmental Microbiology*, **53**: 2282-2295.

Lima dos Santos, C.A.M. (1981). The storage of tropical fish on ice - a review. *Tropical Science*, **23**: 97-127.

Little, C.L., Adams, M.R., Anderson, W.A. and Cole, M.B. (1992). Comparison of a quadratic response surface model and a square root model for predicting the growth rate of *Yersinia enterocolitica.* (Submitted for publication).

Lowry, R.K. and Ratkowsky, D.A. (1983). A note on models for poikilotherm development. *Journal of theoretical Biology*, **105**: 453-459.

Lowry, P.D., Gill, C.O. and Pham, Q.T. (1989). A quantitative method of determining the hygienic efficiency of meat thawing processes. *Food Australia*, **41:** 1080-1082.

Mackey, B .A. and Kerridge, A.L. (1988). The effect of incubation temperature and inoculum size on growth of salmonellae in minced beef. *International Journal of Food Microbiology*, **6**: 57-66.

Maier, V.P., Tappel, A.L. and Volman, D.H. (1955). Reversible inactivation of enzymes at low temperatures. Studies of temperature dependence of phosphatase- and peroxidase-catalyzed reactions. *Journal of the American Chemical Society*, **77**: 1278-1280.

Mallette, M.F. (1969). Evaluation of growth by physical and chemical means. In *Methods in Microbiology*, Volume 1. J.R. Norris and D.W. Ribbons *Eds.* Academic Press, London, pp. 521-566.

Manske, W.J. (1983). The application of controlled fluid migration to temperature limit and time temperature integrators. International Institute of Regrigeration, *Proceedings of the XVIth International Congress of Refrigeration, Paris*, **3**: 397-803.

Matthews, B.W. (1987). Genetic and structural analysis of the protein stability problem. *Biochemistry*, **26**: 6885-6888.

Maxcy, R.B. and Liewen, M.B. (1989). Function of preliminary incubation of raw milk samples for quality control. *Journal of Dairy Science*, **72**: 1443-1445.

McCullagh, P. and Nelder, J.A. (1989). *Generalized Linear Models.* 2nd Edition. Chapman and Hall, London.

McMeekin, T.A. (1988). Preliminary observations on psychrotrophic and psychrophilic, heterotrophic bacteria from Antarctic water samples. *Hydrobiologia*, **165**: 35-40.

McMeekin, T.A. and Franzmann, P.D. (1988). Effect of temperature on the growth rate of halotolerant and halophilic bacteria isolated from Antarctic saline lakes. *Polar Biology*, **8**: 281-285.

McMeekin, T.A. and Olley, J. (1986). Predictive microbiology. *Food Technology in Australia*, **38**: 331-334.

McMeekin, T.A., Chandler, R.E., Doe, P.E., Garland, C.D., Olley, J., Putro, S. and Ratkowsky, D.A. (1987). Model for the combined effect of temperature and water activity on the growth rate of *Staphylococcus xylosus*. *Journal of Applied Bacteriology,* **62**: 543-550.

McMeekin, T.A., Olley, J. and Ratkowsky, D.A. (1988). Temperature effects on bacterial growth rates. In *Physiological Models in Microbiology*. M.J. Bazin and J.I. Prosser *Eds*. CRC Press Inc., Boca Raton, Florida, pp. 75-89.

McMeekin, T.A., Ratkowsky, D.A., Olley, J. and Ross, T. (1989). Comparison of the Schoolfield (non-linear Arrhenius) model and the Square Root model for predicting bacterial growth in foods - a reply to C. Adair *et al. Food Microbiology*, **6**: 304-308.

McMeekin, T.A., Ross, T., and Olley, J. (1992). Application of predictive microbiology to assure the quality and safety of fish and fish products. *International Journal of Food Microbiology*, **15**: 13-32.

Miki, H. and Hamada, T. (1977). Fundamental studies on the thawing of fish. II. Kinetics and prediction on skipjack meat discolouration. *Memoirs of the Faculty of Fisheries, Kagoshima University*, **26**: 129-135.

Miki, H. and Nishimoto, J. (1984). Kinetic parameters of freshness-lowering and discolouration based on temperature dependence in fish muscles. *Bulletin of the Japanese Society of Scientific Fisheries*, **50**: 281-285.

Mistry, V.V. and Kosikowski, F.V. (1983). Use of time-temperature indicators as quality control devices for market milk. *Journal of Food Protection*, **46**: 52-57.

Mohr, P.W. and Krawiec, S. (1980). Temperature characteristics and Arrhenius plots for nominal psychrophiles, mesophiles and thermophiles. *Journal of General Microbiology*, **121**: 311-317.

Murphy, K.P., Privalov, P.L. and Gill, S.J. (1990). Common features of protein unfolding and dissolution of hydrophobic compounds. *Science*, **247**: 559-561.

Neidhardt, F.C., Ingraham, J.L. and Schaechter, M. (1990). The effects of temperature pressure and pH. Chapter 8 in *Physiology of the Bacterial Cell: A Molecular Approach*, Sinauer Associates Inc., Publishers, Sunderland, Massachusetts, pp. 226-246.

Nelder, J.A. and Wedderburn, R.W.M. (1972). Generalized linear models. *Journal of the Royal Statistical Society, Series A*, **135**: 370-384.

Neumeyer, K. (1992). Effect of temperture history on predicting the growth response of *Staphylococcus aureus*. BSc(Hons) Thesis, University of Tasmania.

Nixon, P.A. (1971). Temperature integration as a means of assessing storage conditions. In *Report on Quality in Fish Products*, Seminar No. 3, Fishing Industry Board, Wellington, New Zealand, pp. 34-44.

Nixon, P.A. (1975). New Zealand Patent No. 176879.

Nixon, P.A. (1977). Temperature function integrator. United States Patent 4,061,033.

Ohta, F. and Hirahara, T. (1977). Rate of degradation in cool-stored carp muscle. *Memoirs of the Faculty of Fisheries, Kagoshima University*, **26**: 97-102.

Ohye, D.F. and Christian, J.H.B. (1967). Combined effects of temperature, pH and water activity on growth and toxin production by *Clostridium botulinum* types A, B and E. In *Botulism* 1966. *Proceedings of the 5th International Symposium on Food Microbiology*, pp. 217-223.

Ohye, D.F., Christian, J.H.B. and Scott, W.J. (1967). Influence of temperature on the water relations of growth of *Clostridium botulinum* type E, In *Botulism* 1966. *Proceedings of the 5th International Symposium on Food Microbiology*, pp. 136-143.

Olley, J. (1971). Handling of abalone. In *Report on Quality of Fish Products,* Seminar No. 3, Fishing Industry Board, Wellington, New Zealand, pp. 89-95.

Olley, J. (1978). Current status of the theory of the application of temperature indicators, temperature integrators, and temperature function integrators to the food spoilage chain. *International Journal of Refrigeration,* **1**: 81-86.

Olley, J. (1983). Temperature effects on histamine producing bacteria. In *Properties and Processing of Marine Foods*, National Taiwan College of Marine Science and Technology, Marine Food Science Series. B.S. Pan *Ed.* First Printing Limited Company, Keelung, Taiwan, pp. 14-36.

Olley, J. (1986). In defence of library budgets: A tribute to two librarians. *Food Technology in Australia*, **38**: 276-280, 289.

Olley, J. and Ratkowsky, D.A. (1973a). Temperature function integration and its importance in the storage and distribution of flesh foods above the freezing point. *Food Technology in Australia*, **25**: 66-73.

Olley, J. and Ratkowsky, D.A. (1973b). The role of temperature function integration in monitoring of fish spoilage. *Food Technology in New Zealand*, **8**: 13, 15, 17.

Olley, J. and Quarmby, A.R. (1981). Spoilage of fish from Hong Kong at different storage temperatures. 3. *Tropical Science*, **23**: 147-153.

Owen, D. and Nesbitt, M. (1984). A versatile time-temperature function integrator. *Laboratory Practice*, **33**: 70-75.

Palumbo, S.A., Williams, A.C., Buchanan, R.L. and Phillips, J.G. (1991). Model for the aerobic growth of *Aeromonas hydrophila* K144. *Journal of Food Protection*, **54**: 429-435.

Pfeil, W. and Privalov, P.L. (1979). Conformational changes in proteins. In *Biochemical Thermodynamics*, M.N. Jones *Ed.* Elsevier Scientific Publishing Company, New York, pp. 75-115.

Phillips, J.D. and Griffiths, M.W. (1987). The relation between temperature and growth of bacteria in dairy products. *Food Microbiology*, **4**: 173-185.

Pooni, G.S. and Mead, G.C. (1984). Prospective use of temperature function integration for predicting the shelf-life of non-frozen poultry-meat products. *Food Microbiology*, **1**: 67-68.

Privalov, P.L. and Gill, S.J. (1988). Stability of protein structure and hydrophobic interaction. In *Advances in Protein Chemistry*, **39**: 191-234.

Prusik, T.(1990). Freshness Indicators: A tool for monitoring shelf-life and expanding distribution of fresh seafood products. In *Advances in Fisheries Technology and Biotechnology for Increased Profitability*. M.N. Voigt and J.R. Botta *Eds*. Technomic Publishing Company Inc., Lancaster, PA (USA), pp. 225-233.

Ratkowsky, D.A. (1983). *Nonlinear Regression Modelling: A Unified Practical Approach*, Marcel Dekker, New York.

Ratkowsky, D.A. (1986). A suitable parameterization of the Michaelis-Menten enzyme reaction. *Biochemical Journal*, **240**: 357-360.

Ratkowsky, D.A. (1990). *Handbook of Nonlinear Regression Models*, Marcel Dekker, New York.

Ratkowsky, D.A., Olley, J., McMeekin, T.A. and Ball, A. (1982). Relationship between temperature and growth rate of bacterial cultures. *Journal of Bacteriology*, **149**: 1-5.

Ratkowsky, D.A., Lowry, R.K., McMeekin, T.A., Stokes, A.N. and Chandler, R.E. (1983). Model for bacterial culture growth rate throughout the entire biokinetic temperature range. *Journal of Bacteriology*, **154**: 1222-1226.

Ratkowsky, D.A., Ross, T., McMeekin, T.A. and Olley, J. (1991). Comparison of Arrhenius-type and Bělehrádek-type models for prediction of bacterial growth in foods. *Journal of Applied Bacteriology*, **71**: 452-459.

Reichel, M.P., Phillips, D.M., Jones, R. and Gill, C.O. (1991). Assessment of the hygienic adequacy of a commercial hot boning process for beef by a temperature function integration technique. *International Journal of Food Microbiology*, **14**: 27-42.

Reichardt, W. and Morita, R.Y. (1982). Temperature characteristics of psychrotrophic and psychrophilic bacteria. *Journal of General Microbiology*, **128**: 565-568.

Richards, F.M. (1991). The protein folding problem. *Scientific American*, January: 34-41.

Richards, J.C.S., Jason, A.C., Hobbs, G., Gibson, D.M. and Christie, R.H. (1978). Electronic measurement of bacterial growth. *Journal of Physics. E: Scientific Instruments*, **11**: 560-568.

Roberts, T.A. (1990). Predictive modelling of microbial growth. *Food Technology International, Europe*, pp. 231-235.

Roberts, T.A., Gibson, A.M. and Robinson, A. (1981a). Factors controlling the growth of *Clostridium botulinum* types A and B in pasteurised cured meats. I. In pork slurry in the pH range 5.54 - 6.36. *Journal of Food Technology*, **16**: 239-226.

Roberts, T.A., Gibson, A.M. and Robinson, A. (1981b). Factors controlling the growth of *Clostridium botulinum* types A and B in pasteurised cured meats. II. In pork slurry in the pH range 6.27 - 6.72. *Journal of Food Technology*, **16**: 267-281.

Roberts, T.A., Gibson, A.M. and Robinson, A. (1981c). Prediction of toxin production by *Clostridium botulinum* in pasteurised pork slurry. *Journal of Food Technology*, **16**: 337-355.

Robinson, R.A. and Stokes, R.H. (1949). Tables of osmotic and activity coefficients of electrolytes in aqueous solution at 25°C. *Transactions of the Faraday Society*, **45**: 612-624.

Ronsivalli, L.J. and Charm, S.E. (1975). Spoilage and shelf life prediction of refrigerated fish. *Marine Fisheries Review*, **37**: 32-34.

Ross, T. (1987). Bělehrádek temperature functions and growth of organisms. *CSIRO-DSIR Joint Workshop on Seafood Processing*, Nelson, New Zealand, April 1986. CSIRO Tasmanian Regional Laboratory Occasional Paper No. 18. (Available from CSIRO Marine Laboratory, Castray Esplanade, Hobart, Tasmania.)

Ross, T. and McMeekin, T.A. (1991). Predictive microbiology: Applications of a square root model. *Food Australia*, **43**: 202-207.

Russell, N.J. (1990). Cold adaptation of microorganisms. *Philosophical Transactions of the Royal Society, London. B*, **326**: 595-611.

SAS (1987). *SAS/STAT Guide for Personal Computers*, Version 6 Edition. SAS Institute Inc., SAS Circle, Box 8000, Cary, North Carolina, 27512-800, U.S.A.

Sasajima, M. (1973). Studies on psychrotrophic bacteria in fish and shellfish. 4. Relation between the number of trimethylamine oxide reducing bacteria and their activity. *Bulletin of the Japanese Society of Scientific Fisheries*, **39**: 515-518.

Sasajima, M. (1974). Studies on psychrotrophic bacteria in fish and shellfish. 5. The growth or viability of trimethylamine reducing psychrotrophic bacteria and their activity at subzero temperatures. *Bulletin of the Japanese Society of Scientific of Fisheries*, **40**: 625-630.

Schlatter, D., Kriech, O., Suter, F., and Zuber, H. (1987). The primary structure of the psychrophilic lactate dehydrogenase from *Bacillus psyschrosaccharolyticus*. *Biological Chemistry*, **368**: 1435-1446.

Schoolfield, R.M., Sharpe, P.J.H. and Magnuson, C.E. (1981). Non-linear regression of biological temperature-dependent rate models based on absolute reaction-rate theory. *Journal of Theoretical Biology*, **88**: 719-731.

Schnute, J. (1981). A versatile growth function with statistically stable parameters. *Canadian Journal of Fish and Aquatic Science*, **38**: 1128-1140.

Scott, W.J. (1936). The growth of micro-organisms on ox muscle. I. The influence of water content of substrate on rate of growth at -1°C. *Journal of the Council of Scientific and Industrial Research, Australia,* **9:** 177-190.

Scott, W.J. (1937). The growth of micro-organisms on ox muscle. II. The influence of temperature. *Journal of the Council of Scientific and Industrial Research, Australia,* **10:** 338-350.

Scott, W.J. (1957). Water relations of food spoilage microorganisms. *Advances in Food Research,* **7**: 83-127.

Scott, T. and Brewer, M. (Editors) (1983). *Concise Encyclopaedia of Biochemistry.* (Translation of: Brockhaus ABC Biochemie, H.D. Jakubke and H. Jeschkeit *Eds.*) Walter de Gruyter & Co., Berlin.

Scott, W.J. and Vickery, J.R. (1939). Investigation on chilled beef. Part II. Cooling and storage in the meat works. *Council of Scientific and Industrial Research, Australia,* Bulletin No. 129.

Scott, D.N., Fletcher, G.C. and Hogg, M.G. (1986). Storage of snapper fillets in modified atmospheres at -1°C. *Food Technology in Australia*, **38**: 234-238.

Sharpe, P.J.H. and DeMichele, D.W. (1977). Reaction kinetics and poikilotherm development. *Journal of Theoretical Biology*, **64**: 649-670.

Shaw, M.K., Marr, A.G. and Ingraham, J.L. (1971). Determination of the minimal temperature for growth of *Escherichia coli. Journal of Bacteriology*, **105** : 683-684.

Shechter, E., Letellien, L. and Therisod, H. (1978). Functional *lac* carrier proteins in cytoplasmic membrane vesicles isolated from *Escherichia coli*. In *The Proton and Calcium Pumps*, G.F. Azzone, M. Avron, J.C. Metcalfe, E. Quagliarello and N. Siliprandi *Eds*. Elsevier/North Holland Biomedical Press, pp. 137-159.

Sherlock, M., Fu, B., Taoukis, P.S. and Labuza, T.P. (1991). A systematic evaluation of time-temperature indicators for use as consumer tags. *Journal of Food Protection*, **54**: 885-889.

Shewan, J.M., MacIntosh, R.G., Tucker, C.G. and Ehrenberg, A.S.C. (1953). The development of a numerical scoring system for the sensory assessment of the spoilage of wet white fish stored in ice. *Journal of the Science of Food and Agriculture,* **4**: 283-298

Smith, M.G. (1985). The generation time, lag time and minimum temperature of growth of coliform organisms on meat, and the implications for codes of practice in abattoirs. *Journal of Hygiene, Cambridge*, **94**: 289-300.

Smith, M.G. (1987). Calculation of the expected increases of coliform organisms, *Escherichia coli* and *Salmonella typhimurium*, in raw blended mutton tissue. *Epidemiological Infections,* **99**: 323-331.

Snedecor, G.W. and Cochran, W.G. (1967). *Statistical Methods*. 6th Edition. Ames, Iowa: The Iowa State University Press.

Socket, P.N. (1991). The economic implications of human salmonella infection. *Journal of Applied Bacteriology*, **71**: 289-295.

Spencer, R. and Baines, C.R. (1964). The effect of temperature on the spoilage of wet fish. I. Storage at constant temperature between -1°C and 25°C. *Food Technology, Champaign,* **18**: 769-772.

Sperber, W.H. (1983). Influence of water activity on food-borne bacteria - a review. *Journal of Food Protection,* **46**: 142-150.

S-PLUS. Statistical Sciences Inc., P.O. Box 85625, Seattle, Washington, 98145-1625, USA.

Sprugel, D.G. (1983). Correcting for bias in log-transformed allometric equations. *Ecology*, **64**: 209-210.

Stanier, R.Y., Doudoroff, and Adelberg, E.A. (1971). *General Microbiology*. The Macmillan Press Ltd., London.

Stannard, C.J., Williams, A.P. and Gibbs, P.A. (1985). Temperature-growth relationships for psychrotrophic food spoilage bacteria. *Food Microbiology*, **2**: 115-122.

Stetter, K.O., Fiala, G., Huber, R., Huber, A., Neuner, A. and Segerer, A. (1989). Hyperthermophilic bacterial communities within terrestrial and marine hydrothermal areas. In *Recent Advances in Microbial Ecology*. T. Hattori, Y. Ishida, Y. Marijama, R.Y. Morita and A. Uchida *Eds*. Japan Scientific Services.

Stewart, I. (1990). *Does God play dice, the new mathematics of chaos*, Penguin Books, London.

Storey, R.M. and Owen, D. (1982). A simple time-temperature function integrator. *Unpublished.*

Sturtevant, J.M. (1977). Heat capacity and entropy changes in processes involving proteins. *Proceedings of the National Academy of Science, U.S.A.*, **74**: 2236-2240.

Sydenham, P.H. (1985). *Transducers in Measurement and Control*, 3rd Edition. Adam Hilger Limited, Bristol.

Taoukis, P.S. and Labuza, T.P. (1989a). Applicability of time-temperature indicators as shelf life monitors of food products. *Journal of Food Science*, **54**: 783-788.

Taoukis, P.S. and Labuza, T.P. (1989b). Reliability of time-temperature indicators as food quality monitors under nonisothermal conditions. *Journal of Food Science*, **54**: 789-792.

Taoukis, P.S., Labuza, T.P. and Francis, R.C. (1991). Time-temperature indicators as food quality monitors. *Food Packaging Technology*, ASTM STP 1113, D. Henyon (*Ed.*). American Society for Testing and Materials, Philadelphia, pp. 51-63.

Troller, J.A. (1987). Adaptation and growth of microorganisms in environments with reduced water activity. In *Water Activity: Theory and Applications to Food.* L.B. Rockland and L.R. Beuchat *Eds*. Marcel Dekker Inc., New York, pp. 101-118.

Troller, J.A. and Christian, J.H.B. (1978). *Water Activity and Food,* Academic Press, New York.

Usher, M.J. (1988). *Sensors and Transducers,* MacMillan Publishers Limited, Houndmills.

van Dam, K., Mulder. M. M., Teixeira de Mattis and Westerhoff, H.V. (1988). A thermodynamic view of bacterial growth. In *Physiological Models in Microbiology.* M.J. Bazin and J.I. Prosser *Eds*. CRC Press Inc., Boca Raton, Florida, pp. 25-48.

Van't Hoff, J.H. (1884). *Etudes de dynamique chemique*. Frederik Muller Comp., Amsterdam.

Voyer, R. (*in press*). MKES Tools: A microbial kinetics expert system for assessing and developing food production systems. *Proceedings of the International Workshop on the Application of Predictive Microbiology and Computer Modeling Techniques to the Food Industry, Tampa, Florida, April 1992*. Sponsored by Society for Industrial Microbiology.

Walker, S.J. and Jones, J.E. (1992). Predictive microbiology: data and model bases. *Food Technology International-Europe, 1992*: 209-211.

Walker, S.J., Archer, P. and Banks, J.G. (1990). Growth of *Listeria monocytogenes* at refrigeration temperatures. *Journal of Applied Bacteriology*, **68**: 157-162.

Wells, J.H. and Singh, R.P. (1988a). Application of time-temperature indicators in monitoring changes in quality attributes of perishable and semi-perishable foods. *Journal of Food Science*, **53**: 148-153.

Wells, J.H. and Singh, R.P. (1988b). Response characteristics of full-history time-temperature indicators suitable for perishable food handling. *Journal of Food Processing and Preservation*, **12**: 207-218.

Wheeler, K.A., Hocking, A.D., Pitt, J.I. and Anggawati, A.M. (1986). Fungi associated with Indonesian dried fish. *Food Microbiology*, **3**: 351-357.

WHO Expanded Program on Immunization. (1987). *Training Module on the Use of the Vaccine Cold Chain Monitor*. Logistics and Cold Chain for Primary Health Care. No. 27.EPI/Log/84/27 Rev. 1. WHO, Geneva.

Wodzinski, R.J. and Frazier, W.C. (1961). Moisture requirements of bacteria. II. Influence of temperature, pH and malate concentration on requirements of *Aerobacter aerogenes*. *Journal of Bacteriology*, **81**: 353-58.

Zall, R., Chen, J. and Fields, S.C. (1986). Evaluation of automated time-temperature monitoring system in measuring freshness of UHT milk. *Dairy and Food Sanitation*, **6**: 285-290.

Zuber, H. (1988). Temperature adaptation of lactate dehydrogenases. Structural, functional and genetic aspects. *Biophysical Chemistry*, **9**: 171-179.

Zwietering, M.H., de Koos, J.T., Hasenack, B.E., de Wit, J.C. and van't Riet, K. (1991). Modeling of bacterial growth as a function of temperature. *Applied and Environmental Microbiology* **57**: 1094-1101.

Zwietering, M.H., Jongenburger, I., Rombouts, F.M. and van't Riet, K. (1990). Modelling of the bacterial growth curve. *Applied and Environmental Microbiology*, **56**: 1875-1881.

Zwietering, M.H., Rombouts, F.M. and van't Riet, K. (*in press*). Modelling bacterial growth in food production and distribution chains. *Proceedings of the International Workshop on the Application of Predictive Microbiology and Computer Modeling Techniques to the Food Industry, Tampa, Florida, April 1992*. Sponsored by Society for Industrial Microbiology.

AUTHOR INDEX

SUBJECT INDEX

A

B

C

D

D (cont.)

E

F

G

H

I

K

L

M

N

O

P

R

S

T

V

W